2nd International Meeting for Researchers in Materials and Plasma Technology 2013

Journal of Physics: Conference Series Volume 466

Bucaramanga, Colombia
27 February - 2 March 2013

Editors:

Ely Dannier V-Nino

ISBN: 978-1-62993-445-7
ISSN: 1742-6588

Printed from e-media with permission by:

Curran Associates, Inc.
57 Morehouse Lane
Red Hook, NY 12571

proceedings.com

Some format issues inherent in the e-media version may also appear in this print version.

Printed by Curran Associates, Inc. (2014)

For permission requests, please contact the Institute of Physics
at the address below.

Institute of Physics
Dirac House, Temple Back
Bristol BS1 6BE UK

Phone: 44 1 17 929 7481
Fax: 44 1 17 920 0979

techtracking@iop.org

Additional copies of this publication are available from:

Curran Associates, Inc.
57 Morehouse Lane
Red Hook, NY 12571 USA
Phone: 845-758-0400
Fax: 845-758-2634
Email: curran@proceedings.com
Web: www.proceedings.com

TABLE OF CONTENTS

Journal of Physics: Conference Series **466** (2013) 011001 doi:10.1088/1742-6596/466/1/011001

Preface

These proceedings present the written contributions of the participants of the 2nd International Meeting for Researchers in Materials and Plasma Technology, 2nd IMRMPT, which was held from February 27 to March 2, 2013 at the Pontificia Bolivariana Bucaramanga-UPB and Santander and Industrial - UIS Universities, Bucaramanga, Colombia, organized by research groups from GINTEP-UPB, FITEK-UIS. The IMRMPT, was the second version of biennial meetings that began in 2011.

The three-day scientific program of the 2nd IMRMPT consisted in 14 Magisterial Conferences, 42 Oral Presentations and 48 Poster Presentations, with the participation of undergraduate and graduate students, professors, researchers and entrepreneurs from Colombia, Russia, France, Venezuela, Brazil, Uruguay, Argentina, Peru, Mexico, United States, among others. Moreover, the objective of IMRMPT was to bring together national and international researchers in order to establish scientific cooperation in the field of materials science and plasma technology; introduce new techniques of surface treatment of materials to improve properties of metals in terms of the deterioration due to corrosion, hydrogen embrittlement, abrasion, hardness, among others; and establish cooperation agreements between universities and industry.

The topics covered in the 2nd IMRMPT include New Materials, Surface Physics, Laser and Hybrid Processes, Characterization of Materials, Thin Films and Nanomaterials, Surface Hardening Processes, Wear and Corrosion / Oxidation, Modeling, Simulation and Diagnostics, Plasma Applications and Technologies, Biomedical Coatings and Surface Treatments, Non Destructive Evaluation and Online Process Control, Surface Modification (Ion Implantation, Ion Nitriding, PVD, CVD). The editors hope that those interested in the are of materials science and plasma technology, enjoy the reading that reflect a wide range of topics.

It is a pleasure to thank the sponsors and all the participants and contributors for making possible this international meeting of researchers.

It should be noted that the event organized by UIS and UPB universities, through their research groups FITEK and GINTEP, was a very significant contribution to the national and international scientific community, achieving the interaction of different research groups from academia and business sector.

On behalf of the research groups GINTEP - UPB and FITEK - UIS, we greatly appreciate the support provided by the Sponsors, who allowed to continue with the dream of research.

Ely Dannier V-Niño
The Editor

Organizing Committee
MSc Ely Dannier V-Niño (Chairman)
Dr Valeriy D Dugar-Zhabon – UIS
Dr José J Barba Ortega – UNAL
Dr Aduljay Remolina Millán – UPB
Dra Claudia P González Cuervo – UPB
MSc Fernando Durán Flórez – UPB
MSc José L Herrera Suárez – UPB
MSc Uriel Mora Cruz – UPB
MSc Fredy F Parada Becerra – UIS, UPB
MSc Carlos J Jaimes Ochoa – UTS, UPB
Ing Luis C Caicedo Barrera – UPB

National Scientific Committee
Dr Héctor J Dulce Moreno – UFPS
Dr Henry Riascos Landázuri – UTP
Dr Arturo Plata Gómez – UIS
Dr William Gutierrez Niño – UIS
Dr Rafael Á Torres Amaiz – UIS
Dr Jorge H Martínez Téllez – UIS
Dr Darío Y Peña Ballesteros – UIS
Dr Fredy A Rodriguez Prada – UIS
Dr Antonio C Gutíerrez Piñeres – UTB
Dr Alejandro D Martínez Amariz – UIS
Dra Miryam Rincón Joya – UNAL
Dra Zandra Y Lizarazo Mejía – UIS
Dra María F Serrano Guzmán – UPB
Dr (c) Rafael Cabanzo Hernández – UIS
Dr (c) Jesús David González Acosta – UM
MSc Miguel Á Silva Plata – ITSDZ, UIS
MSc Hermes Barrera Ávila – UPB
MSc (c) Hernán A Garnica Villamizar – CIC, UIS

International Scientific Committee
Dr Amado Cabo – Argentina
Dr Eduardo Méndez – Uruguay
Dr Piotr A Tsygankov – Rusia
Dr Humberto J Jiménez – Venezuela
Dr Luis C Ardila Téllez – España
Dra Linda Gil de Fuentes – Venezuela
Dr Dilson Silva dos Santos – Brasil
Dr Edmanuel E Torres Amaríz – Canada
Dr Enrique Mariano Castrodeza – Brasil
Dra (c) Sugheis Liscano – Venezuela
Dr (c) Wilson Vesga Rivera – Inglaterra
Dra (c) Mariana Pereyra Pérez – Uruguay
Dr (c) Manuel J Andrade Romero – Brasil
MSc Elias Esteban Milano – Uruguay

Journal of Physics: Conference Series **466** (2013) 011001 doi:10.1088/1742-6596/466/1/011001

Sponsors

Universidad Pontificia Bolivariana – UPB
Universidad Industrial de Santander – UIS
Unidades Tecnológicas de Santander – UTS
Banco de la República de Colombia
Canaán Multiservicios
S & S Ingnieria SAS
Vortex Company
Intercovamex
Afilasol
Asomecsa
Metal Green
J & G Soluciones Industriales
Grupo de Investigación en Tecnología del Plasma – GINTEP
Grupo de Investigación en Física y Tecnología del Plasma y Corrosión – FITEK

Journal of Physics: Conference Series **466** (2013) 011001 doi:10.1088/1742-6596/466/1/011001

Universidad Pontificia Bolivariana
SECCIONAL BUCARAMANGA

Universidad Industrial de Santander
UIS

2do ENCUENTRO INTERNACIONAL DE INVESTIGADORES EN MATERIALES & TECNOLOGÍA DEL PLASMA

Peer review statement

All papers published in this volume of *Journal of Physics: Conference Series* have been peer reviewed through processes administered by the proceedings Editors. Reviews were conducted by expert referees to the professional and scientific standards expected of a proceedings journal published by IOP Publishing.

Journal of Physics: Conference Series **466** (2013) 012001 doi:10.1088/1742-6596/466/1/012001

Modifying the growth morphology of aluminum crystals by magnetic mirror in a thermal plasma reactor

G Torrente[1], J Puerta[2], N Labrador[3] and A Molina[3]
[1] Mechanical Department, Simon Bolivar University, Caracas, Venezuela.
[2] Physic Department, Simon Bolivar University, Caracas, Venezuela.
[3] Material Science Department, Simon Bolivar University, Caracas, Venezuela.

E-mail: gtorrente@usb.ve

Abstract. Effect of magnetic fields on growth morphology of aluminum crystals was studied in a fluidized bed thermal plasma reactor assisted by magnetic mirrors. Aluminum crystals were precipitated in the reactor using aluminum powder or aluminum-graphite mixture as precursors. The absent of magnetic field was also studied for comparison. Products were characterized by scanning electron microscopy (SEM) and X-ray Diffraction (XRD). Results indicated that, regardless the precursor used, it was observed the presence of aluminum nanowires when the external magnetic mirror was applied, suggesting that magnetic fields are able to modify growth morphology at nanoscale.

1. Introduction

Control and effects of electromagnetic fields on crystals growth morphology and crystalline structure during the synthesis of new species, have been intensely studied in the last thirty years, and the subject has emerged even more with the recent impulse of nanostructured materials.

In 2002, Khomutov *et al.* [1] studied the effect of electric and magnetic fields on the growth of colloidal amorphous particles. They found that the size and shape of synthesized nanoparticles dramatically changed when an external magnetic field was applied, due to the alignment of growth direction with field lines, resulting in needle-like particles.

Similar results were obtained by Kaneko et al. [2], which presented studies on the influence of variation of magnetic fields intensities on the formation of carbon nanotubes using a rf glow-discharge plasma. They established that nanotubes synthesized under the highest intensity magnetic field studied tended to be smaller in diameter and extended with the typical crystalline structure of carbon nanotubes.

Furthermore, Li Shuqin et al. [3] observed preferential growth of silicon nitride (Si_3N_4) under the effect of magnetic fields. Also, L. Bárdoš et al. [4] reported in 2005 the modification of growth morphology of TiN coatings by applying magnetic fields.

In 2006, Takahashi et al. [4] studied the maglev synthesis of new materials using a furnace assisted by magnetic field. They concluded that the synthesis in magnetic fields generates a magnetic orientation effect in anisotropic susceptible materials.

In addition, other studies [5, 6] suggest that magnetic field applied during synthesis of materials, not only affects size and morphology of particles, but also produces an increment on the synthesis efficiency and product quality.

Thus, the objective of this study is to determine if the presence of external magnetic fields influences the growth morphology of aluminum in a thermal plasma reactor fluidized bed.

2. Experimental procedure and results

The fluidized bed thermal plasma reactor was fabricated in the lab, and the external magnetic field was applied in order to determine its effect during precipitation of aluminum crystals. External magnetic consisted of a copper solenoid powered by an HP 13215A power supply of 52.2 Volt DC and 2.7 ampere. The solenoid is rolled on a refrigerated steel reel with 30 mm of internal diameter. The reel generates such magnetization that magnetic flux is more intense at the ends. This configuration is known as a magnetic mirror. In the lower base of the solenoid, it was adapted the thermal plasma torch Plazjet^{TM} 105/15. The solenoid tightly goes into the cylindrical reactor, within which there is an expansion chamber of 200 mm of diameter, where it is expected that liquids and/or vapor condense rapidly and the synthesis of new materials takes place.

The torch was refrigerated by the heat exchanger HE 200 USA using ethylene-glycol at 75 psi and was powered by a power source Plazjet Whitco of 70 ampere and 200 V DC. This energy is used to make the plasma from 13.2 lpm of high purity nitrogen.

Some tests were performed using aluminum powder (Riedel-de Haën # 11010) as a precursor. However, other tests were executed with a mixture aluminum-graphite. The precursor was transported with nitrogen to 1 psi and a feeding rate of 3 g/min. Feed rate was kept constant and controlled by a Sylko Mark XII dosifier (see figure 1).

Figure 1. Experimental set.

The magnetic field was measured experimentally with a gaussmeter YOKOGAWA model 325i. A maximum intensity of 0.1 Tesla (1000 Gauss) was obtained at the ends of the solenoid. For comparison, procedure was applied with and without application of external magnetic field [7].

Products were characterized by scanning electron microscopy and X-ray diffraction using a JEOL/EO 1.0 SEM and an X-ray diffractometer Philips XpertPro, respectively.

3. Results and discussion

Figures 2 and 3 shows the SEM images of aluminum particles precipitated from a fluidized bed thermal plasma reactor using as a precursor aluminum powder (2) or a mixture of aluminum-graphite powder (3); without application of magnetic field 2 (a) and 3 (a) and assisted by magnetic mirrors 2 (b) and 3 (b). It is clear that, regardless the precursor used, there is a morphological difference of aluminum crystals obtained under the magnetic field, respect to those achieved in the absent of the field. It is evident an elongated morphology similar to nanowires with diameter of 27 nm approximately, when compare with micrometric particles obtained when magnetic field is not applied.

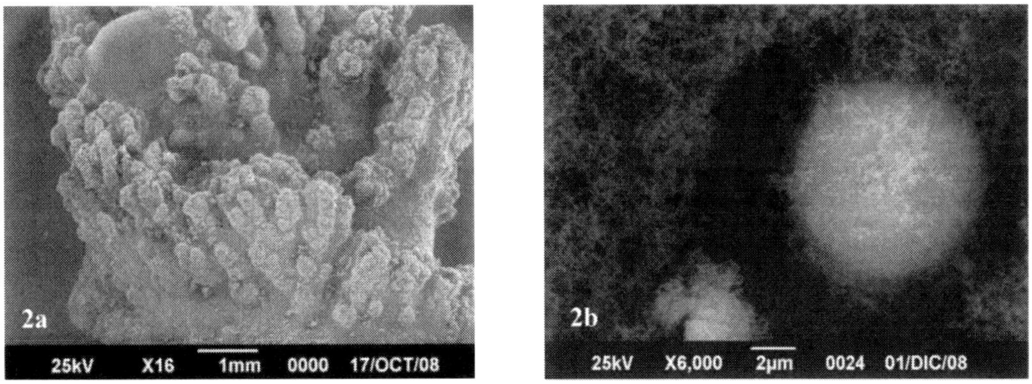

Figure 2. SEM photomicrographs of samples obtained from an assay performed in a thermal plasma reactor employing as aluminum powder precursor without application of external magnetic field 2 (a) and assisted by magnetic mirror 2 (b).

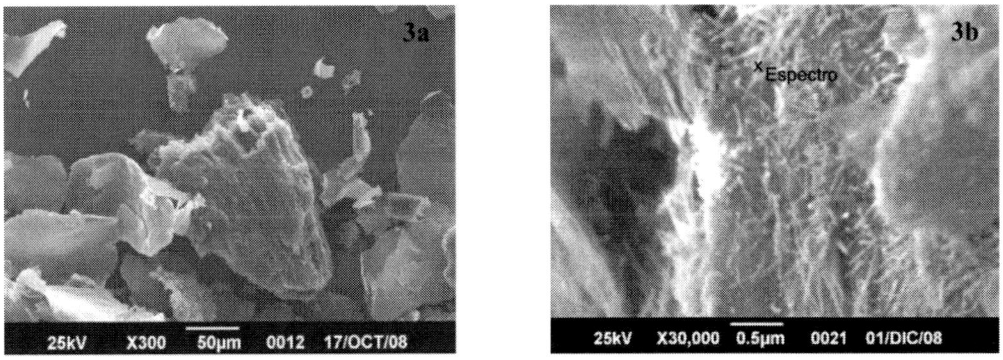

Figure 3. SEM photomicrographs of sample obtained from a test performed in a thermal plasma reactor using as precursor a mixture of aluminum powder with graphite without external magnetic field 3(a) and assisted by magnetic mirror 3 (b).

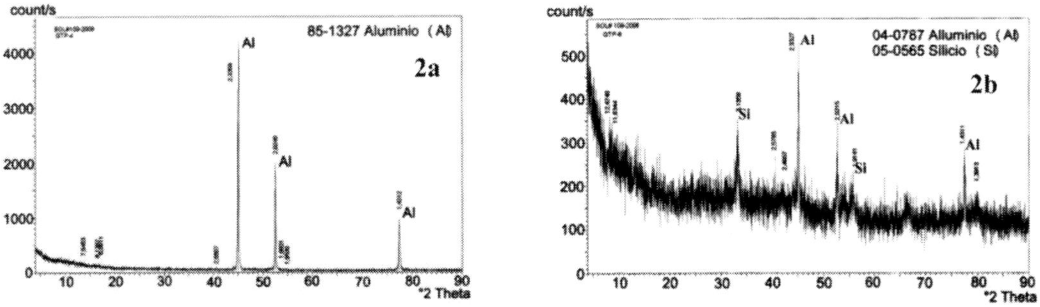

Figure 4. X-ray Diffraction of dendrite synthesized into a thermal plasma reactor and nanotubes synthesized with aluminum powder precursor without magnetic field 4 (a) and assisted magnetic mirror 4 (b).

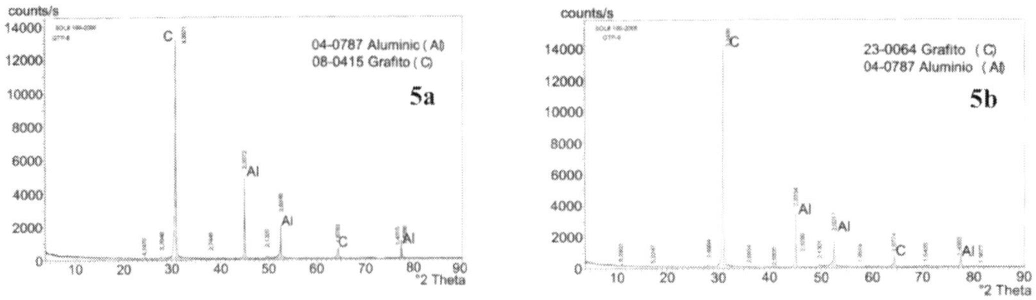

Figure 5. X-ray Diffraction synthesized from a mixture of graphite powder and aluminum precursor in a thermal plasma reactor without magnetic field 5 (a) and with assisted magnetic mirror 5 (b).

Diffractograms depicted in figure 4 and 5 support the results shown previously, due to there is a notably reduction on peak intensities of products achieved under the effect of external magnetic field, typical of nanometric particles. In addition, other diffraction peaks are presented on the spectrum suggesting that magnetic field induces the particle growth through certain preferential direction, in agreement with suggestions of several authors [7].

4. Conclusions

Applying an external magnetic field to a fluidized bed thermal plasma reactor promotes the precipitation of aluminum nanowires, this suggest that the magnetic mirror modifies the crystal growth pattern.

References

[1] Kaneko T, Matsuoka H, Hirata T, Hatakeyama R, Tohji K 2006 *Thin Solid Films* **506-507** 259-62
[2] Doherty S, Buchholz D, Chang R 2005 *Carbon* (Article in Press, Corrected Proof doi:10.1016/j.carbon.12.013)
[3] Li S, Sassa K, Asai S 2008 *Ceram. Intern.* (Article in Press,doi:10.1016/j.ceramint.2005.05.016)
[4] Lee H, Kim Y, Jung W, Han J, Kim J, Bae E 2005 *Surf. and Coat. Tech.* **193** 303-08
[5] Bárdoš L, Gustavsson E, Baránková H 2005 *Surf. and Coat. Tech.* **200** 1862-66
[6] Kohki T, Mogi I, Onogi T, Awaji S, Motokawa M, Watanabe K 2006 *Sci. and Tech. of Adv. Mater.* **7** 346–49
[7] Torrente G, Puerta J, Labrador N 2008 *Phys. Scr.* **014010** (doi:10.1088/0031-8949/2008/T131/014010)

Journal of Physics: Conference Series **466** (2013) 012002 doi:10.1088/1742-6596/466/1/012002

Differences between thin films deposition systems in the production transition metal nitride

J H Quintero[1], A Mariño[2] and P J Arango[3]
[1] Materiales Nanoestructurados y Biomodelación. Universidad de Medellín, Medellín, Colombia.
[2] Laboratorio de Superconductividad y Nuevos Materiales. Universidad Nacional de Colombia, Bogotá, Colombia.
[3]Laboratorio de Física del Plasma. Universidad Nacional de Colombia, Manizales, Colombia.

E-mail: jhquintero@udem.edu.co

Abstract. The progress in vacuum technology have enabled the development of advanced coatings processes such as plasma assisted systems, which can produce thin films of different composition and optimum properties, that cannot be collected for the same material. The techniques of Pulsed Arc, Ionic Implantation and Sputtering have differences to produce coatings. Currently, AuN films have been grown by different techniques such as ion implantation, Reactive Ion Sputtering and Pulsed Arc, which have differences in the grown of the film. Siller 2002 reported a binding energy of 396.6 eV to N1s narrow spectrum as the first direct observation of a gold nitride. In this work, AuN thin films were grown in a system Plasma-Assisted Physical Vapor Deposition by pulsed arc technique. A N1s spectra was obtained with binding energies of 398.1, which by means of the differences between the techniques of ion implantation, sputtering and pulsed arc is concluded have been assigned to gold nitride species.

1. Introduction

The physics of the electrical discharges study the process of the generation, maintenance and transport of electrical currents in an ionized gas. Between 1950 -1960 it started the real development of new plasma technologies in relation with materials process and thin films [1]. I. Langmiur in early 1920 introducing the word "Plasma" to describe the ionized gas, for through of yours studies on the valve to allow the pass of high currents. He defined this term as a gas of neutral and charged particles (cuasi-neutrals) that present collective behavior. The "Plasma" is related with the possibility of "give form" using electrical and magnetics fields, thus avoiding the need of a container as the fluids, liquids or gases. The plasmas are generated through high energetic process, as are the electric discharges in gases or the heating to high temperatures of a gas [1]. The luminescence discharges (electrical discharge type) developed in ionized gas are produced through an electrical discharge to low pressure and are used to thin films production, because in this zone is produced a high of ion bombardment on cathode and the relation current-voltage is almost lineal [2]. The material treatment from plasmas generated for electrical discharges are classified in: i) deposited thin films to low and atmospheric pressure (< 10 - 20 Torr); ii) process of plasma attacks to low pressure (< 0.05 Torr) and different types of superficial treatments to atmospheric pressures; iii) use plasma as ion source, that are used subsequently in the surface treatment. This paper is a recollection of the different techniques used to produce AuN (Transition Metallic Nitride), Ionic Implantation, Sputtering Reactive and Arc Pulsed.

The differences in the plasma generation to produce a thin film influence in the nitride specie obtained, which in turn have its own chemical state, physical and mechanics properties [3]. N1s spectra narrow were obtained for the X-Ray Photoelectron Spectroscopy (XPS) to AuN samples growth for arc pulsed technique and compared with N1s narrow spectrum obtained for Siller to nitride species of gold for Sputering reactive and Impantation Ionic techniques [4, 5], and ends with analysis on the influence of the techniques in the production of nitrides species of gold.

2. Experimental

AuN thin films growth for arc pulsed technique were obtained in the Plasma Physics Laboratory (LAFIP) of the Universidad Nacional de Colombia [6]. This technique was used due to the high ionization grade in the moment of the discharge, which would lead to obtain species to more high binding energies of the N1s Narrow spectrum that showed Siller with Ionic Implantation and Sputtering Reactive. The arc used in this system is of "Cathode Arc", in which two electrodes are face to face, the position of cathode is gold and the anode is stainless steel. The process to obtaining the thin films, the chamber was evacuated to a vacuum of about 10^{-7} mbar, the capacitor bank was charged at 200 V power, alongside the substrate was heated to a temperature of 140 ° C. nitride pressure. The nitrogen pressure in the discharge time for each sample was varied in a range of 2.0 to 8.0 mbar; 3 discharges were performed with the same parameters and then nitrogen was introduced until a pressure of 10^{2} mbar, to hold the films deposited for 3 hr. For comparison with the other techniques, only used M3 (3.5 mbar), which introduced species nitride gold. To obtain spectra N1s narrow a His 165 and AXIS ULTRA, with resolution of 0.35 eV was used. Lorenzianas-Gaussian functions and a Shirley background were used to obtain the binding energies.

3. Results and discussion

In the figure 1 is showed the N1s narrow spectrum to M3, M7 and M10 samples obtained for arc pulsed technique. The absence of nitrogen in M10 could be attributed to the decreased by the mean free path of the particles within the plasma since this was produced at a pressure 7.0 mbar. Para M10 no fue posible la deconvolución del espectro angosto N1s, debido a la baja intensidad del pico (indicativo de baja concentración de N). En M3 se encontró hibridación del espectro angosto N1s en el rango de 396.0 a 398.0 ±0.2 eV (398.1 ±0.2 eV), the which have been attributed to species nitrogen with transition metals [5, 7, 8]. The highest peaks frequently appear in the XPS binding energy analysis to samples grown in plasma assisted systems, due to the vacuum that is used at the time of discharge (~ 1 mbar) and are attributed to the formation of carbo-nitrides and oxy-nitrides [9, 10]. N1s narrow spectra obtained for samples grown by ion implantation, reported 396.7 ± 0.1 eV and 397.7 ± 0.1 eV binding energies as gold nitrided species [4], further observed that the heat treatment changes the narrow spectral line N1s spectrum, which has been attributed to differences in the thermal stability of the nitride species. For samples produced by sputtering were reported 397.5 ± 0.1 and 398.9 ± 0.1 eV binding energies, the low energy was attributed to chemical bonds between gold and nitrogen [5] (also observed by ion implantation). The highest energy peak was previously assigned to oxy-nitride bonded with gold [8, 11]. The difference between the three binding energies in the N1s narrow spectrum can be explained by means of the acceleration of the particles arriving at the substrate; the Ion Implantation have an acceleration few eV [1], the Sputtering have between 10 to 40 eV [1], and the Pulsed Arc technique have acceleration from 50 to 150 eV [12, 13, 14], this difference is influenced by the properties of the plasma produced [1, 15]. The high average energy of the particles (Te) in nonthermal plasmas (usually used in these systems), makes it possible in the collision processes including the gas atoms and molecules are activated by high energy chemical reactions leading to the formation of free radicals in physicochemical environments where the temperature of neutral species (Tg) is reduced 300-1000 K. In this collisions, is produced too ionized species and, above all, to the formation of highly excited atoms and molecules. In the early stages of thin film growth occur a series of steps at the microscopic level, with its own complex mechanism: arrival of the surface atoms (arrival energy of the atoms), subsequent adsorption and surface diffusion (growth rate), nucleation, formation of nuclei and growth of a continuous layer (pressure and temperature). The set of all these stages largely determines the structure and morphology of the deposited layer and therefore its physicochemical properties [16]. The atoms that reach the surface with high energy (so that they have more time to thermalization with the surface) correspond to a higher crystalline structure. Conversely, when the arrival energy of the atoms is small the lenght of diffusion is smaller, and the positions of adsorption would reach lower binding energy.

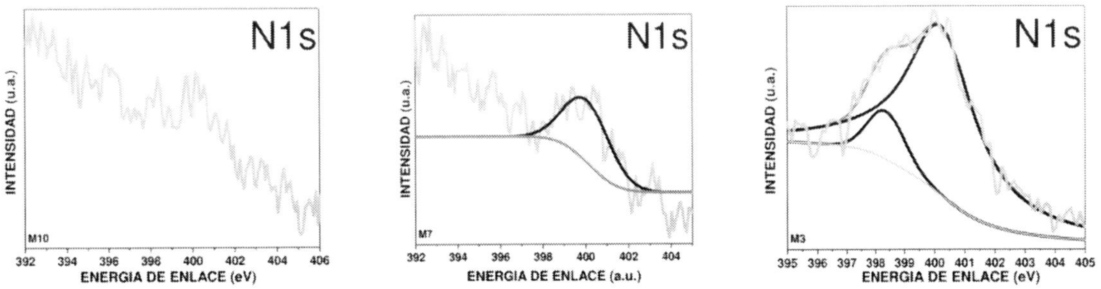

Figure 1. N1s narrow Spectrum, to Samples growth for Arc Pulsed.

Once the islands coalesce and form a continuous layer of material deposited on the substrate, the atoms arrive at the surface and are incorporated to microstructure of the coating, through process of the superficial diffusion and absorption (As in the initial stages); also is show processes like diffusion adatoms along grain joints, to fill the pores and channels which are formed in the film growth; the latter two processes generally appear at higher temperatures and are thermally activated [17].

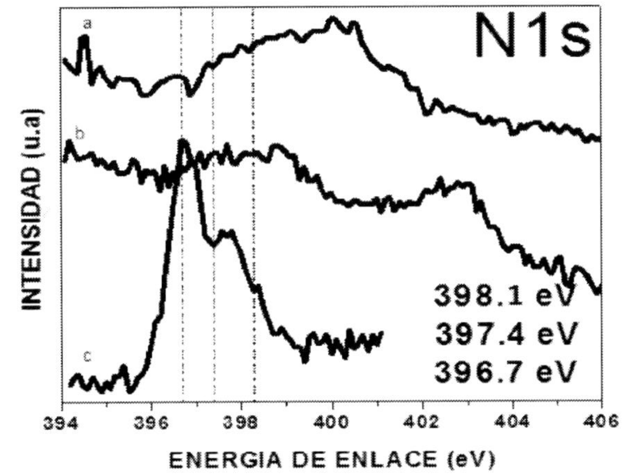

Figure 2. N1s Narrow Spectrum Superposition a) Pulsed Arc, b) Sputtering [6] y c) Ion Implantation [5].

In the spectra obtained for Transition Metallic Nitride, there was an increase of the binding energies (BE), in relation with the system used to growth thin film (see figure 2). The sample of Ion Implantation 396.6 eV, Sputtering 397.5 eV and Pulsed Arc 398.1 eV, which lead to think that the plasma produced in each of this technique, has the appearance of different gold nitride species. The chemical change of a material is directly related to the total charge of the atom (if the load is reduced, increases binding energy). Leading to: i) Increased number of substituents ii) Electronegativity of substituent iii) formal oxidation state. Usually chemical changes are considered as initial state effects.

Chemical changes of $Ti2p_{1/2}$ and $2p_{3/2}$ change to Ti and T^{4+}. The charge then would change Ti → Ti^{4+}, then the 2p orbitals is relax to higher binding energies. Very similarly a change of this type is observed in Transition Metallic Nitride.

4. Conclusions

We compared three production systems Transition metal nitrides (Ion Implantation, Sputtering Reactive and Pulsed Arc) and it was observed that the acceleration of the particles when reaching the surface to have remarkable impact on the production of different species nitride.

Acknowledgments

The authors gratefully acknowledge the financial support of the vicerrectoría de Investigaciones de la Universidad de Medellín under grant 650 and Instituto Colombiano para el Desarrollo de la Ciencia y la Tecnología (COLCIENCIAS) in your program "doctorados nacionales en la II Convocatoria Nacional de Apoyo a Programas de Posgrado 2006".

References

[1] Jose M Albella 2003 *Láminas Delgadas y Recubrimientos Preparación Propiedades y Aplicaciones* (España: Consejo Superior de Investigaciones Científicas)
[2] Jhon Jairo Olaya 2005 *Recubrimientos de Nitruros Metálicos Depositados con la Técnica de Spurreo Asistido con Campos Magnéticos Variables* (México: Universidad Nacional Autónoma de México)
[3] A Pauschitz, M Roy, F Franek 2008 *Tribol. Int.* **41** 584–602
[4] L Siller, M R C Hunt, J W Brown, J M Coquel, P Rudolf 2002 *Surf. Sci.* **513** 78
[5] S Krishnamurthy, M Montalti, M G Wardle, M J Shaw, P R Briddon, K Svensson, M R C Hunt and L Siller 2004 *Phys. Rev.* B. **70** 045414
[6] A Devia, V Benavides, E Restrepo, D F Arias, R Ospina 2006 *Vacuum* **81** 378–84
[7] I Bertoti 2002 *Surface and Coatings Technology* **151** 194 - 203
[8] A Devia, H A Castillo, V J Benavides, Y C Arango, J H Quintero 2008 *Materials Characterization* **59** 105–07
[9] C C Yu, S Ramanathan, F Sherif, S T Oyama 1994 *J. Phys. Chem.* **98** 13038
[10] R Kapoor, S T Oyama, B Fruhberger, J G Chen 1997 *J. Phys. Chem.* B **101** 1543
[11] Yu V Butenko, L Alves, A C Brieva, J Yang, S Krishnamurthy, L Siller 2006 *Chem. Phys. Let.*, **430** 89
[12] Isak Beilis 2001 *IEEE Transactions on Plasma Science* **29-5** 657-70
[13] Raymond L Boxman, Philip J Martin, David Sanders 1995 *Hanbook of vacuum arc science and technology* (New Jersey: Noyes Publication)
[14] Andre Anders 2001 *IEEE Transactions on Plasma Science* **29-2** 393-98
[15] Burkhard Jutnner 2001 *J. phys. D: Applied Phys.* **34** 103-23
[16] Nowakowska-Langier Katarzyna, Chodun Rafal, Nietubyc Robert, Minikayev Roman, Zdunek Krzysztof 2013 *Applied Surface Science* **275** 14-8
[17] David M Sanders, Andre Anders 2000 *Surf and Coat Techno,* **133-134** 78 – 90

Journal of Physics: Conference Series **466** (2013) 012003 doi:10.1088/1742-6596/466/1/012003

Synthesis and evaluation of MgF_2 coatings by chemical conversion on magnesium alloys for producing biodegradable orthopedic implants of temporary use

P Y Casanova[1], K J Jaimes[1], N J Parada[1], C A Hernández-Barrios[2], M Aparicio[3], F Viejo[1] and A E Coy[2]
[1] Escuela de Ingeniería Química, Universidad Industrial de Santander, Bucaramanga, Colombia.
[2] Escuela de Ingeniería Metalúrgica y Ciencia Materiales, Universidad Industrial de Santander, Bucaramanga, Colombia.

E-mail: anaecoy@uis.edu.co

Abstract. The aim of the present work was the synthesis of biodegradable MgF_2 coatings by chemical conversion on the commercial Elektron 21 and AZ91D magnesium alloys, in aqueous HF solutions for different concentrations and temperatures. The chemical composition and morphology of the coatings were analyzed by scanning electron microscopy (SEM-EDX) and X-ray diffraction (XRD). On the other hand, their corrosion behavior was evaluated by gravimetric and electrochemical measurements in Hank´s solution at 37°C for different immersion times. The experimental results revealed that chemical conversion in HF produced MgF_2 coatings which corrosion resistance was enhanced by increasing the HF concentration. Further, the microstructure and composition of the base alloy played a key role on the growth and degradation mechanisms of the MgF_2 coatings.

1. Introduction

Nowadays, research into biomedical applications has been focused on a new category of biodegradable implants [1]. These devices can perform a specific function for a certain period of time and then gradually disappear, avoiding long-term complications, and progressively transfer the load to the surrounding tissue. In these sense, Mg alloys potential candidates as biodegradable materials due to several advantages: Mg is non-toxic and it may stimulate the growth of new bone tissue [2]; its in-vivo degradation generates corrosion products that can easily be dissolved, absorbed and/or excreted after tissue healing, avoiding the need of the second additional surgery to remove the implant [3, 4]; and their mechanical properties (density, elastic modulus, and compressive yield strength) are close to those of the human bone that is a great advantage in order to minimize the stress shielding phenomenon [5].

However, the main limitation of magnesium alloys for their use as implant material is its excessively high degradation rate in body fluids, which can lead to the loss of mechanical properties of the implant before the bone has healed. Further, the corrosion process of magnesium alloys also generates hydrogen gas bubbles that, if the local hydrogen saturation of blood and tissues are exceeded, might accumulate in the tissue cavities [6, 7]. In this regard, there are generally two possible ways to improve the corrosion behaviour of Mg and its alloys: i) tailor their composition and microstructure and ii) modified their electrochemical behaviour through surface treatment. Among the surface treatments currently used for magnesium alloys, the chemical conversion route is one of the most simple and effective processes to produce a protective layer on the surface. Particularly, chemical conversion in hydrofluoric acid allow obtaining magnesium fluoride (MgF_2) coating that has been reported to be an effective alternative for improving the corrosion resistance of potential biodegradable Mg-based implants [8-10]. Based on these statements, the aim of the work was to synthesize and evaluate MgF_2 coatings by chemical conversion on magnesium alloys for producing biodegradable orthopaedic implants for temporary use.

2. Experimental procedure

Two magnesium alloys were used in the present investigation: the AZ91D (Al: 8.5-9.5%; Zn: 0.45-0.9%; Mn: 0.17-0.3%; others: 0.08%; Mg: balance (%wt.)) and the Elektron 21 (Zn: 0.2-0.5%; Nd: 2.6-3.1%; Gd: 1.0-1.7%; Zr: saturated; Mg: balance (%wt.)) magnesium alloys. These alloys were supplied by Magnesium Elektron Ltd. specimens were wet ground through successive grades of silicon carbide abrasive papers, from 120 to 600, rinsed in alcohol and dried in a warm stream of air. For conversion treatment, the samples were immersed vertically in a plastics bottle containing HF of concentration 1.6, 2.4, 3.2 and 4.0 v/v %. at room temperature for 24 h. The treated samples were then rinsed with deionized water, dried at 120°C (30 min.) and weighed. Coatings were characterized using Scanning Electron Microscopy (SEM). Finally, corrosion resistance of the coatings was evaluated by immersion and anodic potentiodynamic polarization tests in Hank´s solution at 37°C.

3. Results and discussion

Figure 1 shows the electron micrographs of the MgF_2 coatings performed by chemical conversion in HF 4%v/v solution at room temperature and 24 hours on the AZ91 and elektron 21 magnesium alloys.

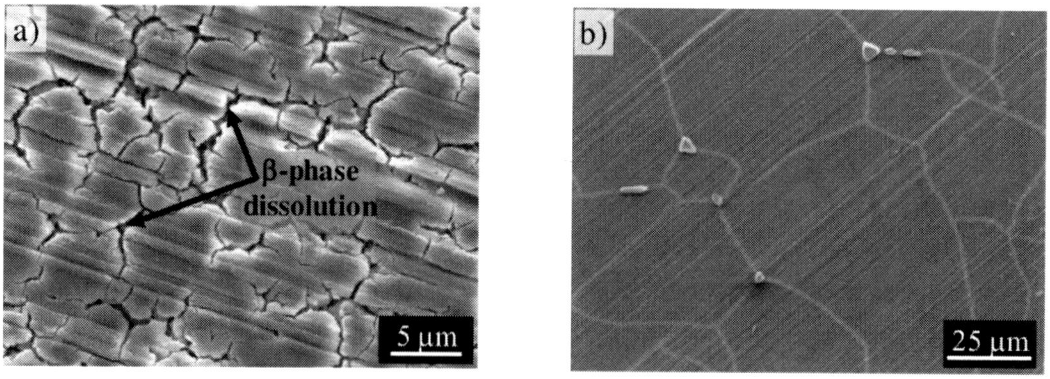

Figure 1. Electron micrographs of the MgF_2 coatings performed by chemical conversion in HF 4%v/v solution at room temperature: (a) AZ91 and (b) elektron 21 magnesium alloys.

The preferential dissolution of the reactive β-phase ($Mg_{17}Al_{12}$) in the AZ91D alloy promotes the growth of irregular MgF_2 coatings with thickness ranged between 3 and 10 µm. Conversely, the more homogeneous microstructure of elektron 21 alloy favoured the formation of continuous, uniform coatings, 5 µm thick, where the intermetallic $Mg_{12}(Nd_xGd_{1-x})$ phase remained stable on the top surface, without any evidence of partial dissolution or degradation.

Figure 2 represents the mass gain versus time for the AZ91D and Elektron 21 alloys treated by chemical conversion in different HF solution after immersion in Hank´s solution.

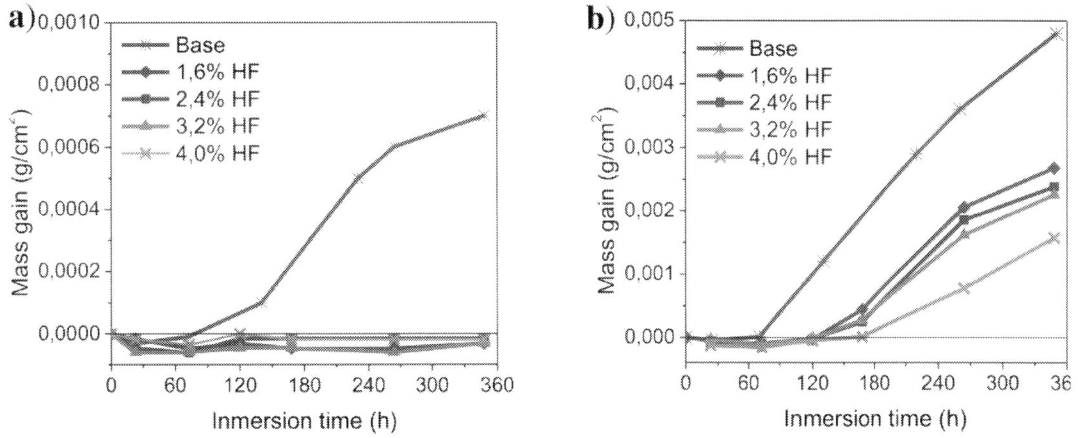

Figure 2. Mass gain curves of the magnesium alloys treated in different HF solutions after immersion in Hank´s solution at 37°C for: (a) AZ91D and (b) elektron 21.

It was noticed that chemical conversion of magnesium alloys significantly reduced the corrosion rate of the specimens, especially for the AZ91D alloy (figure 2 (a)). Interestingly, coatings performed on the Elektron 21 specimens showed a corrosion mechanism divided in two stages or process, in which inflexion point probably indicates the beginning of the dissolution of the magnesium alloy underneath. Moreover, it is also evident that an increase in the HF concentration affected positively the corrosion behavior, retarding the degradation of the metallic substrate. This particular behavior may be suitable in design of biodegradable coatings for short-term applications.

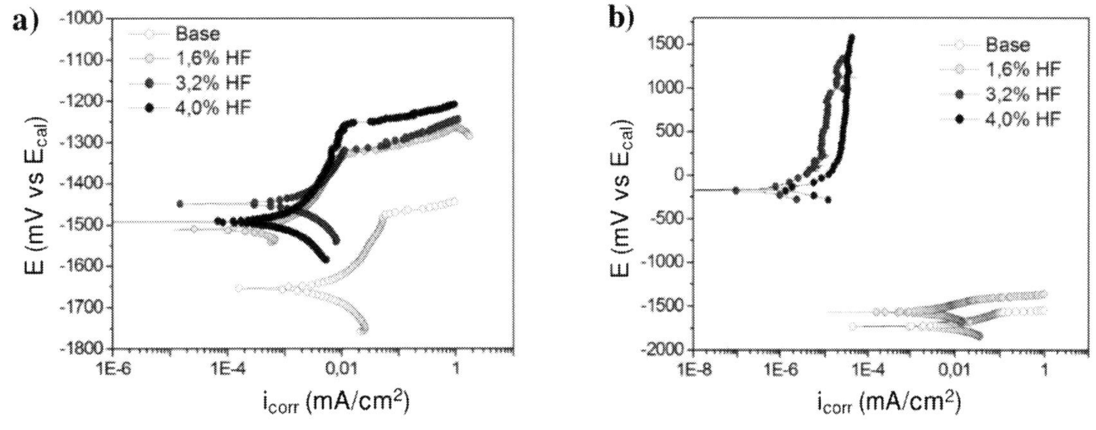

Figure 3. Anodic polarization curves of the magnesium alloys treated in different HF solutions after immersion in Hank´s solution at 37°C for 60 min: (a) AZ91D and (b) elektron 21.

On the other hand, figure 3 shows the anodic polarization curves for the same specimens and conditions. Compared to the parent alloy, the AZ91D coated specimens exhibited better corrosion performance with reduction of corrosion current density by about one order of magnitude, while the corrosion potential was displaced toward more noble values. On the other hand, conversion coating of the Elektron alloy resulted in a much more evident improvement. In this regard, the specimens coated in HF solutions with concentrations higher than 3,0% v/v, showed passive properties for the entire potential range selected; and experienced reduction of the corrosion current density in three and four orders of magnitude with a displacement of the corrosion potential by more than 1000 mV.

4. Conclusions

Conversion coating of the AZ91D and Elektron 21 magnesium alloys in HF solution resulted in MgF_2 layer which provide an excellent protection against corrosion, where the microstructure (presence and nature of different intermetallic phases) plays a key role in its growth. Further, an increase in the HF concentration affected positively in the corrosion behavior of both alloys retarding even more the anodic dissolution of the metallic substrate. Coatings performed on the Elektron 21 exhibited an interesting corrosion mechanism divided in two stages, which is suitable in design of biodegradable coatings for short-term applications.

Acknowledgements

The authors wish to thank to the Vicerrectoría de Investigación y Extension of the Universidad Industrial de Santander, Colombia (Project 5456) for the financial and technical support of the present work.

References

[1] Gastaldi D, Sassi V, Petrini L, Vadani M, Trasatti S and Migliavacca F 2011 *J. Mech. Behav. Biomed.* **4** 352-65
[2] Zreiqat H, Howelett C R, Zannettino A, Evans P, Schulze-Tanzil G, Knabe C and Shakibaei M 2002 *J. Biomed. Mater. Res.* **62** 175-84
[3] Carboneras M, García-Alonso M C, Escudero M L 2011 *Corr. Sci.* **53** 1433-9
[4] Hornberger H, Virtanen S and Boccaccini A R 2012 *Acta Biomat.* **8**, pp. 2442-2455
[5] Staiger M P, Pietak A M, Huadmai J and Dias G 2006 *Biomaterials* **27** 1728-34
[6] Chiu K Y, Woung M H, Cheng F T and man H C 2007 *Suf. and Coating Tech.* **202** 590-8
[7] Witte F, Fischer J, Nellesen J, Crostack H-A, Kaese V, Pisch A, Beckmann F and Windhagen H 2006 *Biomaterials* **27** 1013-8
[8] Tingting Y, Tan L, Xiong D, Zhang B and Yan K 2010 *Mater. Sci. Eng.* C **30** 740–748
[9] Thomann M, Krause C, Agrisani N. Bormann D, Hassel T, Windhagen H and Meyer.Linderberg A et al., 2010, *J. Biomed. Mater. Res.* A **93**-4 1609–19
[10] Carboneras M, Hernández-Alvarado L A, Mireles Y E, Hernández L S, García-Alonso M C and Escudero M L 2010 *Rev. Metal. Madrid* **46-1** 86–92

Journal of Physics: Conference Series **466** (2013) 012004 doi:10.1088/1742-6596/466/1/012004

Preparation and characterization of Fe_2O_3 nanoparticles

M R Joya[1], **J Barón-Jaimez**[2] **and J Barba-Ortega**[1]

[1] Departamento de Física, Universidad Nacional de Colombia, Bogotá, Colombia.

[2] Grupo Plasnamat, Instituto de Investigaciones en Materiales, Universidad Autónoma de México, DF, México.

E-mail: mrinconj@unal.edu.co

Abstract. Nano particles hematite (Fe_2O_3) with good crystallinity were prepared by Pechini method. Hematite (Fe_2O_3) has emerged as a promising photo-electrode material due to its significant light absorption, chemical stability in aqueous environments, and ample abundance. Photoelectrochemical cells offer the ability to convert electromagnetic energy from our largest renewable source, the Sun, to stored chemical energy through the splitting of water into molecular oxygen and hydrogen. The structure and the size of the Fe_2O_3 nanoparticles were analyzed by SEM and XRD. The UV-Vis optical absorption of the samples was also investigated. These methods help to define the obstacles that remain to be surmounted in order to fully exploit the potential of this promising material for solar energy conversion.

1. Introduction

Many industrial dyes are toxic and carcinogenic. The organic dyes present in industrial wastewater often pose significant threats against human health and environmental pollution control. Therefore, it is important to remove organic dyes from wastewater. However, wastewater exhibits stable behavior under harsh conditions and resists biodegradation,making it difficult to remove organic dyes easily [1, 2, 3, 4, 5, 6, 7, 8]. Many wastewater treatment methods have been explored for example the adsorption technique, but this method is now becoming unpopular because it is expensive, and the adsorbent has low recyclability. Photocatalysis is an environment friendly process that utilizes irradiation energy for catalytic reactions. Hence, photocatalytic technology has been widely investigated for applications to the decomposition of pollutants. Researchers are especially interested in developing photocatalysts that can extend the absorption wavelength into the visible-light region [1, 2, 9, 10]. Hematites (Fe_2O_3) are minerals belonging to the group of iron-oxide minerals; they have a hexagonal structure and exhibit paramagnetic behavior. Moreover, they show strong catalytic activity, widely and easily available, and are extremely environment friendly. In particular, hematite may be a promising candidate for visible-light photocatalysis it can absorb visible light, collect up to 45% of solar-spectrum energy, and is one of the cheapest semiconductor materials available [2, 9, 10, 13].In this study, hematite nanoparticles are employed for the removal of Rhodamine B (RhB) dye by photocatalytic decomposition. RhB is nonbiodegradable and extensively used in the industry; therefore, it is selected as the model contaminant .

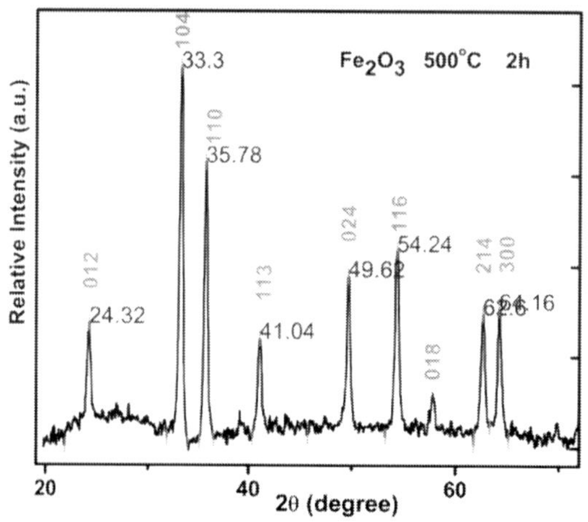

Figure 1. X-ray powder diffraction patterns of Fe_2O_3 prepared by the polymeric precursor method at 500 C.

2. Experimental

The polymeric precursor method (pechini) is based on the polymerization of metallic citrate by ethylene glycol[11, 12]. A hydrocarboxilic acid such as citric acid is normally used to chelate cations in an aqueous solution.The method is based on an intensive blending of positive ions in a solution, controlled transformation of the solution into a polymer gel, removal of the polymer matrix and development of an oxide precursor with a high degree of homogeneity. During the synthetic process, metal salts or alkoxides are introduced into a citric acid solution with ethylene glycol. The formation of citric complexes is believed to balance the difference in individual behavior of ions in solution, which results in a better distribution of ions and prevents the separation of components at later process stages.

3. Results and discussion

The XRD pattern (1) of the annealed sample shows the formation of phase pure hematite.shows the XRD pattern of the dark red-brown materials, Fe_2O_3 obtained after $500C$. The XRD spectrum of the nanoparticles shown in figure (1) contains nine peaks, which are clearly distinguishable [13]. All of them can be perfectly indexed to rhombohedral Fe2O3 in peak position (JCPDS No. 24-0072), space group$R\overline{3}$(148). The spectrum contains no impurity phase. Furthermore, the relative intensity of the (110) diffraction peak is greater than that of polycrystalline Fe_2O_3 power. The fact indicates that the nanoparticles Fe_2O_3 crystals may develop preferentially rather than randomly.

The nanoparticles size was roughly estimated to be about $60 - 70nm$ from the SEM data, as shown in figure 2 Such a type of unique morphology was observed for the first time for iron oxide with excellent magnetic properties. SEM images also clearly showed the nano-structural homogeneities and remarkably unique neck-structured morphology with average size of $50 - 60nm$. Although the neck-structure shaping mechanism is unclear, low temperature during autoclave, long time oxidation and annealing may be responsible to the neck-structure formation. Further study is needed to clarify the mechanism. The figure 2a) shows the sample without addition of NH_4 at a temperature of $500C$ this presents an irregular shape of the nanoparticles, while figure 2 b) is Fe_2O_3 with addition of NH_4 at a temperature of $500C$,these nanoparticles have a more regular basis.

Figure 3 shows the absorption spectra of the RhB solution during different irradiation times.

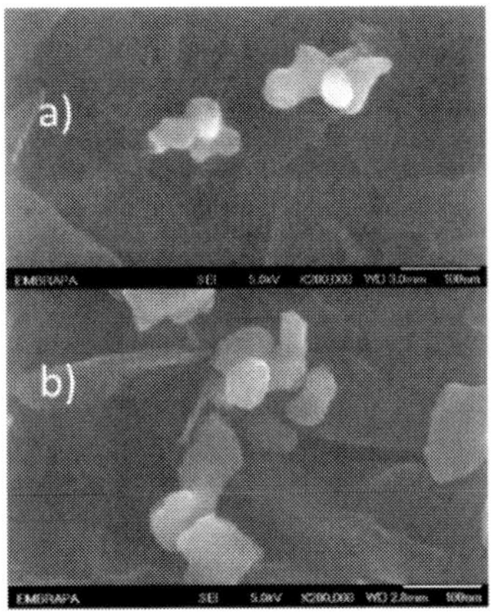

Figure 2. Examples of hematite nanoparticule morphologies: SEM images of hematite prepared by the thermal oxidation method (top image Fe_2O_3, bottom image Fe_2O_3 with NH_4). Temperature $500C$.

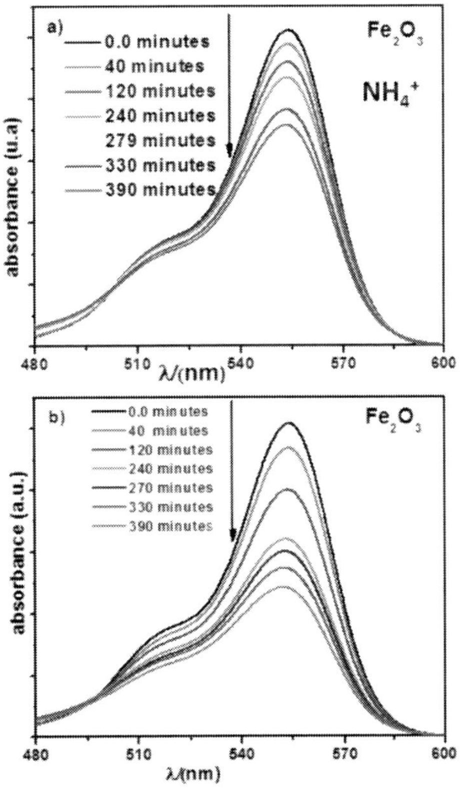

Figure 3. (Color online) Absorption spectra of rhodamine B aqueous solution in presence of nanoparticles of Fe_2O_3 , using $\lambda = 350nm$ for the illumination. a) with addition of NH_4, b) without added NH_4.

In this work, experiments were performed in three different conditions, namely: (1) without catalyst, (2) in the presence of Fe_2O_3 with PH=7 as catalyst and (3) in the presence of

Fe_2O_3 nanoparticles as catalyst. UV irradiation of RhB in the absence of hematite catalyst showed almost a negligible variation in the absorbance maximum. However, when hematite nanostructures were used as photocatalyst, the absorption maximum (at $\lambda = 550$ nm) steadily decreased upon irradiation with UV, without any obvious peak shift of the absorption maximum. This suggests that the complete degradation and decolourization of aqueous RhB was purely due to the catalyst under UV irradiation. This work reports a facile and simple method for the large scale synthesis of hematite nanostructures. The prepared hematite nanostructures showed excellent photocatalytic activity towards Rhodamine B, which was believed to be due to the high surface area. Among these two structures of the hematite, the Fe_2O_3 without NH_4 showed relatively faster photocatalytic activity towards RhB.

Acknowledgements

The authors would like to thank Ely Dannier V-Niño from LNDC-UFRJ and Alejandro Barba for their very useful discussions.

References

[1] S Bharathi, D Nataraj, D Mangalaraj, Y Masuda, K Senthil and K Yong 2010 *Supercond. J. Phys. D* **43** 015501
[2] Yen-Hua and T. Kuo-Jui 2012 *International Journal of Photoenergy* 980595
[3] B Xu, B Huang, H Cheng, Z Wang, X Qin, X Zhang and Y Dai 2012 *Chem. Commun* **48** 6529
[4] X Xue, K Hanna, N Deng 2009 *J. of Hazardous Materials* **166** 407
[5] K Sivula, F L Formal and M Gratzel 2011 *ChemSusChem* **4** 432
[6] H Dotan, K Sivula, M Gratzel A Rothschild and S C Warren 2011 *Energy. Environ. Sci* **4** 958
[7] P S Shinde, G H Go and W J Lee 2012 *J. Mater. Chem* **22** 10469
[8] J Lian, X Duan, J Ma, P Peng, T Kim and W Zheng 2009 *ACS Nano* **11** 3749
[9] L Lu, Z Ai, J Li, Z Zheng, Q Li and L Zhang 2007 *Crystal Growth - Design* **7** 459
[10] Y Liu, L Yu, Y Hu, C Guo, F Zhang and X W Lou 2012 *Nanoscale* **4** 183
[11] D Melo, M Melo, A Martinelli, Z Silva, J Cunha, A Lima 2007 *Ceramica* **53** 79
[12] A A Cavalheiro, C R Foschini, M A Zaghete, C O Paiva-Santos, M Cilense, J A Varela, E Longo 2001 *Ceramics International* **27** 509
[13] Y Fu, J Chen, H Zhang 2001 *Chem. Phys. Lett* **350** 491

Journal of Physics: Conference Series **466** (2013) 012005 doi:10.1088/1742-6596/466/1/012005

Nitrogen implantation into steel wire coated with zinc used as reinforcement in power transmission conductors

J J Castro-Maldonado[1], H J Dulcé-Moreno[1] and E D V-Niño[2,3,4]
[1] GITEC. Universidad Francisco de Paula Santander, Cúcuta, Colombia.
[2] Grupo de Investigación en Tecnología del Plasma - GINTEP, Universidad Pontificia Bolivariana seccional Bucaramanga, Floridablanca, Colombia.
[3] Grupo de Investigación en Física y Tecnología del Plasma y Corrosión - FITEK, Universidad Industrial de Santander, Bucaramanga, Colombia.
[4] Laboratorio de Ensayos No Destructivos, Corrosión y Soldadura – LNDC, COPPE, Universidad Federal de Rio de Janeiro, Rio de Janeiro, Brasil.

E-mail: hjdulce@ufps.edu.co

Abstract. In tropical environments, diversity of climatic factors such as temperature, relative humidity, deposition of environmental contaminants (such as sulfates and chlorides) affect a large proportion of materials exposed to the weather, and electrochemical corrosion is one of the phenomena that occur in the case of metals and alloys [1, 2]. It is therefore particularly important to study this behavior in the Zinc-coated steel, since this material is used for its economy in the industry specifically in the area of transport of electricity.

1. Introduction
Zinc coated steel is an important factor in the structural [3] part of power grids as it supports the efforts produced by the winds and the weight of the same line. The transmission and power distribution lines are basically composed of wire ACSR (aluminum conductor steel reinforced Zinc coated), Since the galvanized Steel wire confers (provides) good corrosion resistance, nevertheless, many cases have been reported in galvanized pipes, which have revealed an eventual corrosion process [4] In the same way, considering the advantages offered by the nitriding process, such as: increased surface hardness and reduction in the corrosion rate due to the formation of nitrides in the steel by the reaction of nitrogen with the iron and alloying elements . Furthermore, with nitriding an increased resistance to embrittlement caused by hydrogen is produced because this material is in a humid environment and in addition supporting mechanical stresses so that the water vapor reacting with fresh surfaces [5] created by fatigue, causes the hydrogen to react with the new surface producing atomic hydrogen diffusion later by placing ahead of the front of the fissure occurring failure [6, 7], the nitrogen thus blocks the interstitial spaces of preventing hydrogen diffusion.

2. Methodology
Samples of Zinc coated steel core wire for ACSR 1/0 trade were selected (see figure 1) which were measured corrosion rate, resting potential and polarization resistance through a potentiostat-galvanostat. The methodology used here is similar to that carried out in Huyuan Sun (e tál) y V. Padilla (et ál) paper (work). [4, 8] Subsequently these samples underwent an implantation nitriding

process and these variables were measured [9]. Additionally, samples nitrided and non-nitrided (see figure 2) were exposed to corrosive environments and mass loss determined gravimetrically.

Figure 1. Standard Cable ACSR.

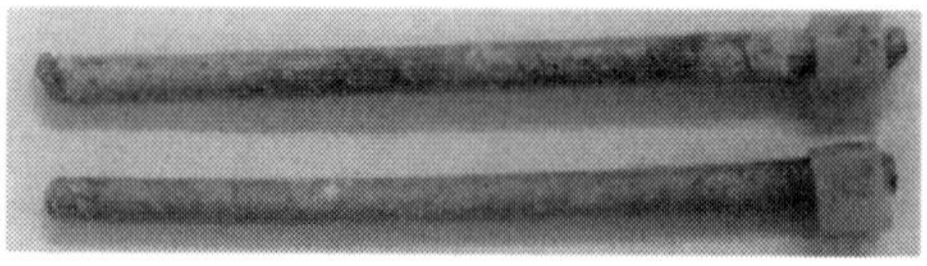

Figure 2. Non-nitrided sample (top) Nitrided sample (bottom).

3. Results and discussion

A potentiometric study was made to each sample through which the corrosion rate measurements given in MPY (mils per year) and the passivation potential were obtained in mV (millivolts). Statistical methods were developed by the respective box and whisker plots where illustrates the variation of the data and the average for each variable of the same shape were arranged the mass loss data obtained by gravimetry (see figure 3). The nitrided samples were first subjected to a cleaning treatment by sputtering with argon gas to 5kV, subsequently to a surface modification 3DII IMPLATATION 20kV with nitrogen gas for 30 minutes (see figure 4) [10].

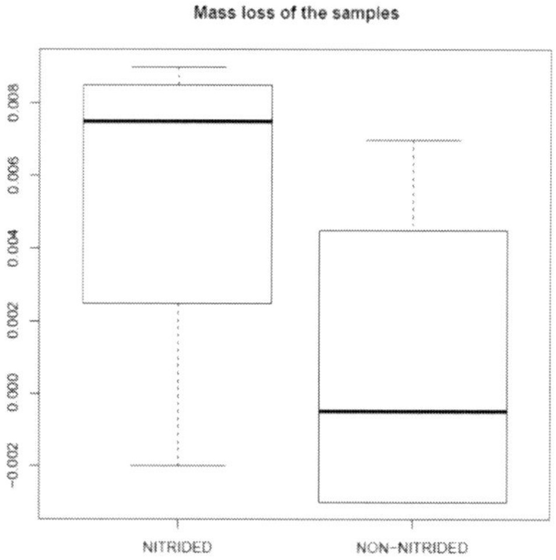

Figure 3. Mass loss of the samples.

Figure 4. 3DII Implantation process.

Based on studies with the potentiostat, specimens of non-nitrided and nitrided, it was evidenced an increase in the corrosion rate of the samples subjected to nitriding treatment by implantation surface, additionally it was noticed that nitriding favors a displacement passivation potential, then passivation phenomenon accelerates in the coating. It is noteworthy that according to Lisbeth Acosta [11] the galvanized zinc has a thickness of about 20 micrometers (see figure. 5). This thickness is insufficient for protecting base steel in aggressive corrosive environments. Moreover, the simulation performed with the code "TRIM" [12] to the nitrogen implantation conditions Zinc made sheds implantation depth of approximately 0.08 micrometers (see figure 6), indicating that the nitriding does not break through the zinc coating, in the event that a nitriding steel core by implantation were required, it must perform before galvanizing.

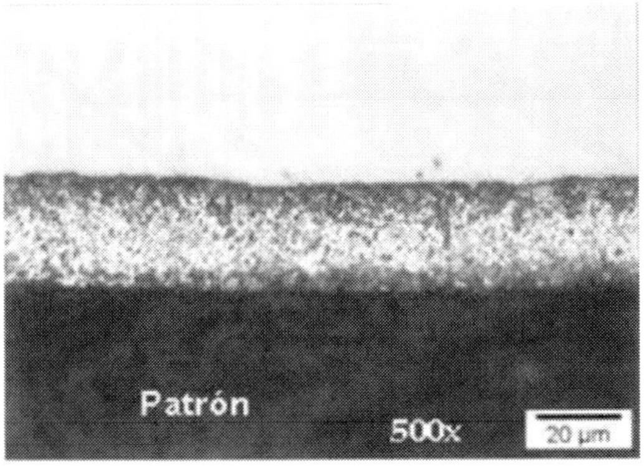

Figure 5. Zinc layer thickness.

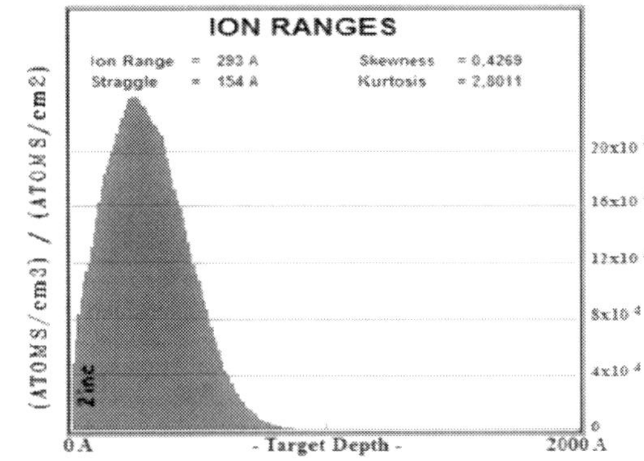

Figure 6. Implantation depth profiles of nitrogen within Zinc.

4. Conclusions

It was found that the nitride implanted galvanized steel samples exhibit a higher corrosion rate than non-nitride ones (samples). It could be due to the fact that nitrogenous implantation depth in zinc (depth in nitrogenous implantation in zinc) is much less than the thickness of the zinc coat in the galvanized steel, therefore Nitrogenous could affect the passivation properties of galvanization (galvanization passivation properties).

References

[1] Molina J D, Villada F 2011 *Revista Facultad de Ingeniería* **59** 257-66
[2] Vera R, Guerrero F, Delgado D, Araya R 2009 *Revista de la Construcción* **8-2** 18-26
[3] Marcolin D 2004 *Memorias Congreso CONAMET/SAM* **2**
[4] Huyuan S, Shuan L, Lijuan S 2013 *Int. J. Electrochem. Sci.* **8** 3494–509
[5] Gonzalez Díaza D, Vargas Cárdenas H 2005 *Revista de la Facultad de Ingeniería y Procesos* 2-6
[6] Dugar-Zhabon V D, Dulcé-Moreno H, Garnica-Villamizar H A, V-Niño E D 2012 *Materials Research* **15-6** 969-73
[7] E D V- Niño, V Dugar-Zhabon, H J Dulce-Moreno, G Peña-Rodríguez, H A Garnica, P A Tsygankov 2012 *Revista ITECKNE* **9-1** 14-20
[8] Padilla V, Alfantazi A 2013 *Corrosion Science* **69-2** 174-85
[9] H J Dulcé-Moreno, P A Tsygankov, V Dugar-Zhabon, E D V-Niño, F F Parada-Becerra 2011 *Puente Revista Científica* **5-2** 21-6
[10] E D V-Niño, J L Pinto, V Dugar-Zhabon and J A Henao 2012 *Journal of Physics: Conference Series* **370** 012031
[11] Acosta L, Castañeda A, Corvo F, Berardo L 2010 *Revista CENIC Ciencias Químicas* **41** 1-11
[12] http://srim.org, 02, 2013

Industrial applications of thermal sprayed coatings in Venezuelan steelmaking industry

S Liscano[1], E Nuñez[2], L Gil[1] and R Zerpa[2]
[1] Departamento de Ingeniería Metalúrgica, Universidad Nacional Experimental Politécnica "Antonio José de Sucre", Puerto Ordaz, Venezuela
[2] Orinoco Iron S.C.S. Technical Management

E-mail: sliscano@unexpo.edu.ve

Abstract. The metal components subjected to high temperature conditions, abrasive wear, corrosion, impact, etc.; tend to present degradation of manufacturing material, causing the failure imminent of the component. One of the alternatives to minimize or eliminate such effect is the application of ceramic coatings, which are thermal insulators and exhibit high mechanical strength. Its extreme hardness, coupled with the low friction properties and chemical stability, allowing its use in a wide variety of applications. Therefore, the following paper describes the application of thermal sprayed coatings obtained by HVOF and Plasma technologies like alternative to protect the metallic equipment in different venezuelan industrial sectors, such as to operate under aggressive conditions of service, such as the steelmaking nationals industries. This study presents applications cases of ceramic-based coatings, in order to minimize the sticking of metallic material in components of reduction reactor of FINMET® and MIDREXTM process.

1. Introduction

In the venezuelan steel industrial park, located on the rivers sides of the Orinoco, adjacent to the mains Venezuelan iron ore mines, there are service conditions with high temperatures, wear and corrosion. These conditions have been studied by Corrosion Research Center of the National Polytechnic University (UNEXPO), which aim to develop protective coatings that reduce failure of metal components, which in turn is consistent with the research policies established by the Venezuelan government in 2011. In Venezuelan steelmaking processes such as direct reduction processes had been found that one of the metallurgical phenomena that can affect the continuity process is the adherence of ore pre-reduced iron ore to metal components of reactors. This condition limits the transfer of material or fluid to the point of stopping operations, as in the case of reactors grills cones in the process FINMET® and the screws of briquetting machines in the MIDREX® process. One alternative to minimize or eliminate this effect is the application of ceramic coatings, which are insulating materials having high mechanical strength, extreme hardness, low friction properties and chemical stability, resulting in a less rough surface which minimizes sticking. Therefore, as described below the application of HVOF and Plasma thermal sprayed coatings represents an alternative to avoid these failures.

2. Direct reduction technologies

A direct reduction process (DR) is used for obtaining metallic iron units, from the chemical interaction between iron ore units and reduction gases such as H_2 and CO obtained through reforming of methane (CH_4) or natural gas, to a process temperature near to 800° C.

2.1. MIDREX^{TM} technology

The Midrex™ Direct Reduction is a process of moving bed shaft furnace where the reducing gas moves counter-current to the lump iron ore or iron ore pellet solids in the bed. The reducing gas (from 10-20% CO and 80-90% H_2) is produced from natural gas using Midrex's CO_2 reforming process and their proprietary catalyst (instead of steam reforming) [1]. The reduced ore is compacted by briquetting machines to minimize re-oxidation. This process requires the feed material to the pockets compaction system through feed screws of HK40 austenitic steels.

2.2. FINMET technology

The Finmet process is a DR of fluidized bed, in which the gas is introduced into train of reactors connected in series, where it flows counter-current contacting the ore from the ore feed systems, like the the above process is reduced material compacted into briquettes [2]. For the fluidization of the ore is necessary to increase the gas velocity, so that a grid is used with perforated carbon steel cones (See figure 1 (b)).

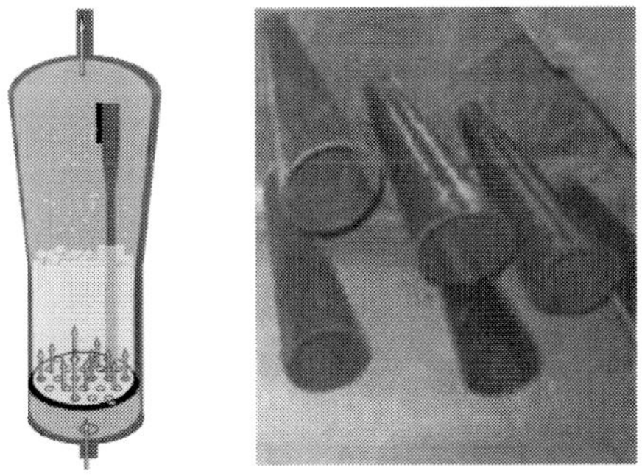

Figure 1. (a) Reactor grill FINMET ®.
(b). Cones grill FINMET ® reactor.

3. Cases thermal spray coating application

To continuation are presented two studies cases in the Venezuelan steelmaking industries.

3.1. Application No. 01

The first case is related to the company: Venezolana de Prerreducidos de Caroní CA (VENPRECAR), which development the MIDREXTM hot briquetted process.

The application was made in the screw briquetting machine feeders, these recurring failures occur sticking reduced fine ore causing plugging and pulling her down the team (See figure 2).

The different parts of the axle-Speyer analyzed indicate that the base metal is effectively formed by the steel HK40, which has a hard coating by welding with a low nickel content (11.43%), causing the formation of ferrite, which is associated with the magnetization of the material, facilitating the formation of a crust [3].

3.2. Application No. 02

The second case is related with the Orinoco Iron S.C.S. that development the FINMET ® fluidized bed Process.

The application was performed on the cones of the R30 reactor grills, which were clogged by the adhering metallic fines which were dragged by the reducer gas from the reactor R20, which increases the pressure differential on the grid affecting the fluidization process inside the reactor and in consequence causing a decreasing of the number of train operating days (See Figure 3). The cones are made with a carbon steel plate A-36 and due to the affinity of them with the metallic fines the sticking are facilitated in these high temperature conditions of the process.

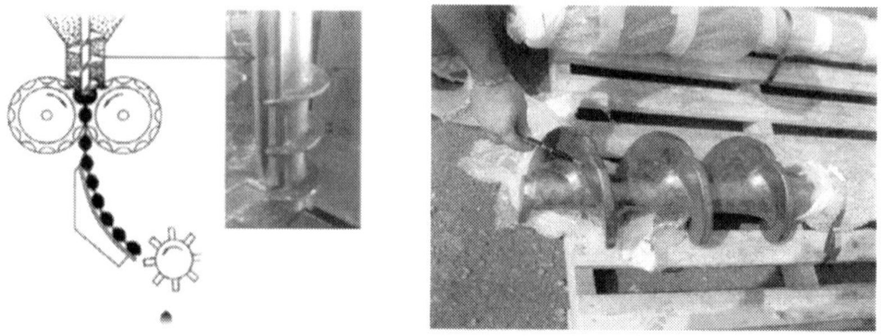

Figure 2. (a) Feeder Screw the briquetting machines. (b) Finishing the product augers with tungsten carbide thermal spray [3].

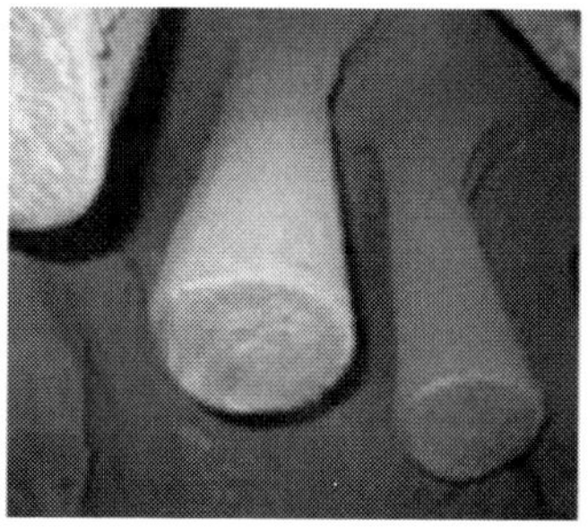

Figure 3. Cones FINMET ® reactor grill clogged.

Tests were performed on a laboratory scale (See figure 4) employing coatings produced by HVOF technology. Coatings were deposited with WC-12%Co [4], and Colmonoy 88 alloy; the last in two conditions: as coated and with flame post- heat treated [4]. The process simulated conditions results indicated that the lowest adherence occurred in the WC-Co coatings exposed to 1 hr to process temperature (figure 5 (a)). Later, coated steel cones were evaluated at industrial scale and they were compared with uncoated cones. The results are showed in the figure 5 (b), where can be distinguishing that the coated cones present minor adherence.

Figure 4. Sticking test equipment.

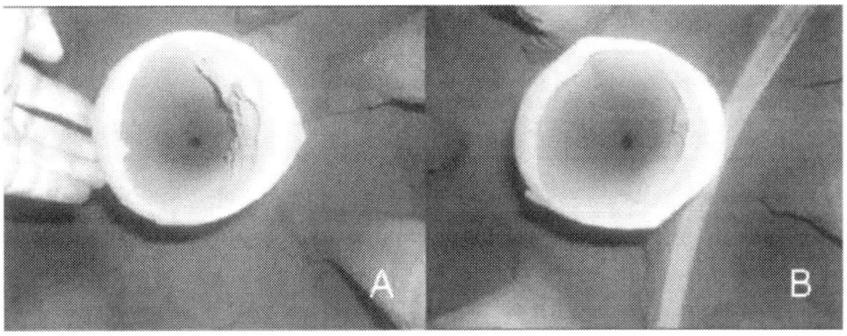

Figure 5. Samples tested with (a) uncoated cone. (b) coated cone.

4. Conclusions and upcoming developments

Although, the applications of thermal sprayed cermets coatings in the steel industry were successful in the present research; however, other alternatives of coatings are very interesting to probe. Thinking in upcoming developments, it's important to highlight that the current trend in development and material selection for high temperature applications has a strong focus on based ceramic oxide coatings, which offer a high resistance to wear, corrosion, hardness, chemical stability at high temperatures and are thermal insulators [5]. Specifically, is necessary to study the applications of thermal barriers better known as TBCs (Thermal Barriers Coatings) for special alloy components. For over 30 years, the plasma sprayed yttria stabilized zirconia (YSZ) coatings, has been used for gas turbine applications, specifically the coatings of yttria partially stabilized (Y-PSZ), which have the metastable tetragonal phase. Oxides (SiO_2, Al_2O_3 and TiO_2) are used as stabilizers which prevent to the yttria suffer a harmful volume expansion that occurs during the phase changes of martensitic type tetragonal to monoclinic [6]. Recent research points to the use of different combinations of stabilizers oxides to improve properties such as thermal conductivity and fracture toughness [7, 8]. Next researches will be focus in optimization of deposition of ZrO_2-TiO_2-Y_2O_3 and Al_2O_3-TiO_2 systems fabricated by atmospheric plasma spray [9].

References

[1] Middle East Steel Industry 2013 *Direct Reduction Process MIDREX* (Venezuela: http://steelmaking.wordpress.com/2008/12/01/direct-reduction-process-midrex/)
[2] Industry Efficiency Technology Database 2013 *Finmet* (Venezuela: http://ietd.iipnetwork.org/content/finmet)
[3] Vásquez Z 2006 *Obstrucción del Tornillo Alimentador del Sistema de Alimentación de las Máquinas Briqueteadoras de Venprecar C.A* (Puerto Ordaz: Unexpo)
[4] S Stewart, R Ahmed, T Itsukaichi 2005 *Surf. and Coat. Tech.* **190** 171
[5] Wang, Dinwiddie R B and D Porter Wallace 2010 *Development of a Thermal Transport Database for Air Plasma Sprayed ZrO2-Y2O3 Thermal Barrier Coatings* (Estates Unites: ASM International)
[6] Mauer, Sebold Doris, Vaßen Robert and Stover Detlev Georg 2009 *Characterization of Plasma-Sprayed Yttria-Stabilized Zirconia Coatings by Cathodoluminescence* (Estates Unites: ASM International)
[7] Schaedler T A, Leckie R M, Krämer S 2007 *Journal of the American Ceramic Society* **90** 12
[8] M Milieska, K Brinkiené, R Kézelis and Césniené J 2009 *Lithuanian Journal of Physics* **49-1** 81
[9] Liscano S 2010 *Correlación de las Características Microestructurales con los Parámetros de Deposición de Recubrimientos de* ZrO_2*-10%Y2O*$_3$*-18%*TiO_2 *Termorrociados por Plasma* (Venezuela: Unexpo-UPM)

The technique of Raman spectroscopy in the characterization physical, chemical and structural ferroelectric of the materials

M R Joya[1], J Barón-Jaimez[2] and J Barba-Ortega[1]

[1] Departamento de Física, Universidad Nacional de Colombia, Bogotá, Colombia.
[2] Grupo Plasnamat, Instituto de Investigaciones en Materiales, Universidad Autónoma de México, DF, México.

E-mail: `mrinconj@unal.edu.co`

Abstract. The ferroelectric ceramics have been one of the most studied compounds in the literature both in obtaining the same growth as in behavior of chemical and physical properties. In its growth the materials can go through other phases or to change their magnetic ferroelectric properties due to vacancies generated as $PbTiO_3$ material. This material structural changes its phase also applying pressure with temperature, concentration and allowing anharmonic study behavior [1]. Raman spectroscopy is a technique which provides high resolution in seconds chemical and structural information of almost any organic or inorganic compound. As a technique which does not require a special atmosphere or vacuum, is one technique that has been used extensively in the characterization of ferroelectric materials in particular of $PbTiO_3$.

1. Introduction

Lead titanate, $PbTiO_3$ (PT), is a known Perovskite kind of ferroelectric ceramic. The substitution of Pb^{2+} to de La^{3+} ions improves the mechanical and ferroelectric characteristics of PT. Due to its potential in technological applications, PLT has been widely studied [1, 2, 3, 4, 5, 6, 7, 8, 9, 10, 11, 12, 13, 14, 15]. The physical properties of PLT depend mainly on the concentration of Lanthanum. It is also known that the particle size, the morphology and the crystal structure govern the ferroelectric structure in this material. PT has a high phase transition temperature from tetragonal ferroelectric to cubic paraelectric ($Tc = 770k$), but it decreases with increasing concentration of Lanthanum. The same thing occurs with hydrostatic pressure applied ($Tc = 12GPa$). These effects were verified in other published papers[16, 17, 15]. The Titanium of (PT) exhibits a negative thermal expansion (NTE), whose temperature depends on the lattice parameters which changes with La-concentration) [18].The phononic modes in PLT have been studied before [2, 4, 5]. However, besides the recent work by Joya and Ahart *et al* [1, 6], an evolution of Raman mode with the temperature and the pressure has not been documented in details.

2. Experimental

Raman scattering measurements were performed using a Jobin-Yvon T64000 triple monochromator with a CCD detector. The exciting light from an argon laser was focused

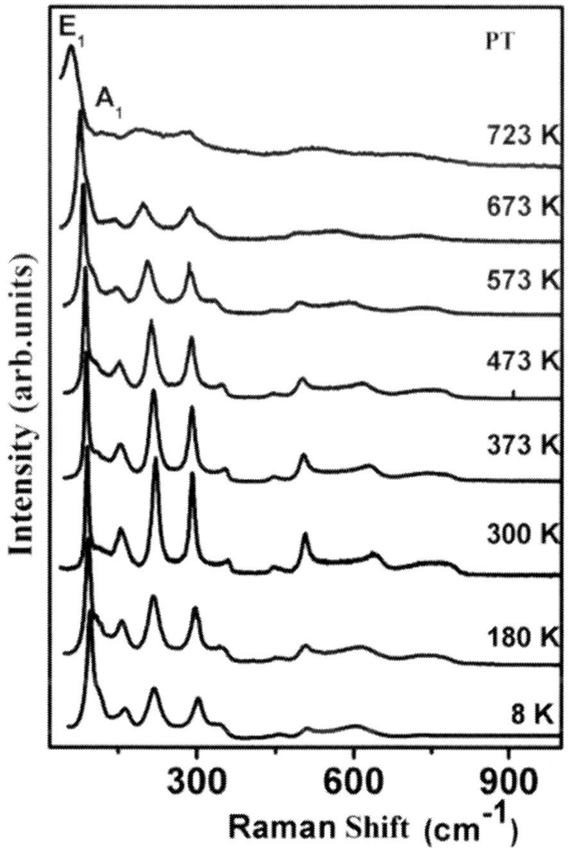

Figure 1. Raman spectra with temperature ($Pb_{1-x}La_xTiO_3$) for samples with La concentration, in this case $x = 2.5\%$.

on the samples with an optical microscope using a long working distance 50X objective. For the low temperature measurements, in the 10 K to 300 K range, a helium closed cycle cryostat was used while for high temperature, in the 300 K to 850 K range, an optical micro furnace was used. High pressure was applied on the samples using a diamond anvil cell (DAC), stainless steel gaskets with 100 micrometers thickness, and holes with a diameter of about 120 micrometers. Within our experimental conditions, we could not reach a pressure above 7.5 GPa, since the holes in the gaskets were deformed. The pressure transmitting medium was methanol-ethanol in the classic proportion 4:1. The frequency values of each vibrational mode were obtained by fitting asymmetric Lorentzian line shapes.

3. Results

In Figure 1 shows the Raman spectra as a function of temperature, normal modes of vibration go towards low frequencies with increasing temperature.The temperature affects the population of the different levels for each normal mode due to the non- equidistance in levels of energy, since the average position of atoms changes. However, in both cases, the change in the network parameters is attributed to the potential anharmonicity. On the other hand, the variations of the half-width of Raman bands under the effect of temperature or pressure are attributed to the decay to the decay of a given phonon into two or three phonons [19].

In figure 2 the softening of all the Raman peaks is due to the increase of chemical and

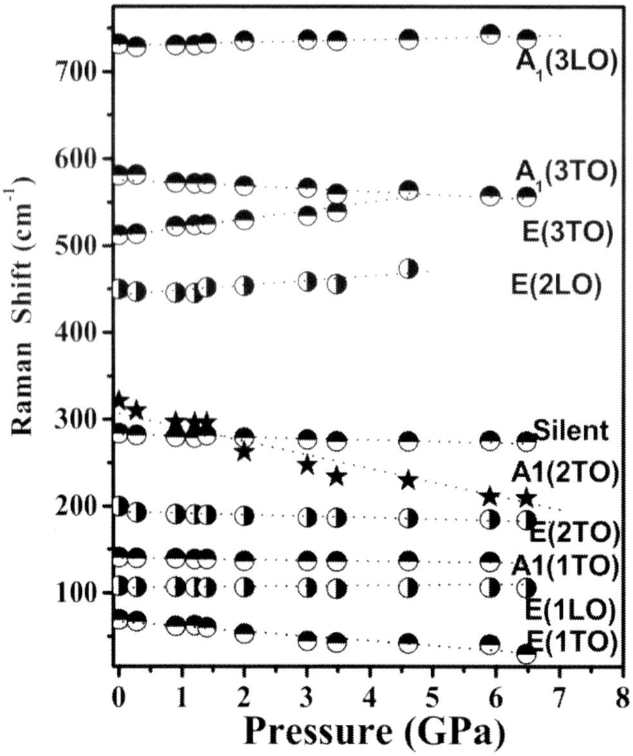

Figure 2. The Raman spectra for the region low frequency for the 2.5% La sample. This figure shows the softening evolution of the E1(TO) mode with increasing applied hydrostatic pressure.

structural disorders. Similar frequency behavior shifts and broadening were observed for the other La concentrations. The peak softening could be due to the defects and dislocations present in the lattice, local symmetry breakdown for the impurities and vacancies. All these effects in the lattice can lead to a breakdown of the Raman selection rules, this effect is observed in the damping of the vibration modes. Where P is the pressure, α the thermal expansion, $d\omega$ the shift of Raman bands, T the temperature, and α the Coefficient of thermal expansion. This equation can be written as a function of bulk modulus [20, 21, 22]:

$$\left(\frac{\partial\omega}{\partial T}\right)_P = -B\alpha\left(\frac{\partial\omega}{\partial P}\right)_T + \left(\frac{\partial\omega}{\partial T}\right)_V \Rightarrow \left(\frac{\partial\omega}{\partial T}\right)_V = B\alpha\left(\frac{\partial\omega}{\partial P}\right)_T + \left(\frac{\partial\omega}{\partial T}\right)_P \qquad (1)$$

The first term $(\frac{\partial\omega}{\partial T})_V$ in equation 1 denotes the phonon-phonon interaction, or is the so-called explicit contribution of anharmonicity. To the right of the equation the volume-dependent implicit term $(\frac{\partial\omega}{\partial P})_T$ can be obtained by measuring the Raman spectrum under pressure at constant temperature. The term $(\frac{\partial\omega}{\partial T})_P$ can be determined by measuring the spectrum as a function of the temperature at constant pressure. The coefficients $(dw/dp)_T$ are related through the coefficient of thermal expansibility and compressibility of the materials.

As can be observed data found in other work [19], with increasing concentration the coefficient also increases to an absolute value. In this case, we were unable to calculate the coefficient for 15 and 20% La concentration, since the bulk modulus values have not been reported in the literature. In table 1 the result of the coefficient of anharmonicity for the pure material is comparable to that of literature obtained by simulation of the molecular dynamic [23].

Table 1. Calculation of $(\frac{\partial\omega}{\partial T})_V$ coefficients at constant volume. Explicit contribution of anharmonicity.

PLT ($E1$)	0% La	2.5%La	5%	10% La
$(\frac{\partial\omega}{\partial T})_V$ (cm^{-1})	-0.039	-0.054	-0.06	-0.08

Acknowledgements

The authors would like to thank Ely Dannier V-Niño from LNDC - UFRJ and A. Barba for their very useful discussions.

References

[1] M R Joya, J Barba-Ortega, P S Pizani, 2013 *J. Appl. Phys* **113** 013512
[2] Z Cai, X Xing, R Yu, G Liu, Q Xing 2006 *J. of Alloys and Compounds* **420** 273
[3] J A Sanjurjo, E López-Cruz, G Burns 1983 *Phys. Rev. B* **28** 7260
[4] Z Kutnjak, J Petzelt and R Blinc 2006 *Nature. Lett* **441** 956
[5] E C Paris, M F C Gurgel, M R Joya, G P Casali, C O Paiva-Santos, T M Boschi, P S Pizani, J A Varela, E Longo 2010 *J. of Phys. and Chem. of sol* **71** 12
[6] M Ahart, M Somayazulu, R E Cohen, P Ganesh, P Dera, H. Mao, R Hemley, Y Ren, P Liermann and Z Wu 2008 *Nature. Lett* **451** 6459
[7] Y Liu, G Xua, C Song, Z Ren, G Han, Y Zheng 2008 *Materials Science and Engineering A* **472** 269
[8] P S Dobal and R S Katiyar 2002 *J. Raman Spectrosc* **33** 405
[9] J Zhu, H Xu, J Zhang, C Jin, L Wang and Y Zhao 2011 *J. Appl. Phys* **110** 084103
[10] S Zhang and F Li 2012 *J. Appl. Phys* **111** 031301
[11] S Zhai, W Xue, C Lin, X Wu, E Ito 2011 *Phys. Chem Minerals* **011** 0436-1
[12] E Simon, F Borodavka, I Gregora, D Nuzhnyy, S Kamba, J Hlinka, A Bartasyte and S Margueron 2011 *J. Appl. Phys.* **110** 084115
[13] T Shimada, Y Uratani and T Kitamura 2012 *Appl. Phys. Lett* **100** 162601
[14] A Crassous, V Garcia, K Bouzehouane, S Fusil, A H G Vlooswijk, G Rispens, B Noheda, M Bibes and A Barthélémy 2010 *Appl. Phys. Lett* **96** 042901
[15] N Jaouen, A C Dhaussy, J P Itie, A Rogalev, S Marinel and Y Joly 2007 *Phys. Rev. B* **75** 224115
[16] M R Joya and P S Pizani 2010 *Appl. phys. Lett* **97** 031903
[17] P Ganesh and R E Cohen 2009 *J. Phys.Condens. Matter* **21** 064225
[18] J Chen, X Xing, R Yu and G Liu 2005 *J. Am. Ceram. Soc* **88** 1356
[19] M R Joya, J Barba-Ortega and P S Pizani 2013 *Jour. App. Phys* **113** 013512
[20] S K Srivastava 2006 *Solid State Sciences* **8** 573
[21] A Bartasyte, S Margueron, J Kreisel, P Bourson, O Chaix-Pluchery, L Rapenne-Homand, J Santiso, C Jimenez, A Abrutis, F Weiss and M D Fontana 2009 *Phys. Rev. B* **79** 104104
[22] I Efthimiopoulos, K Kunc, S Karmakar, K Syassen, M Hanfland and G Vajenine 2010 *Phys. Rev. B* **82** 134125
[23] S Costa, P Pizani, J Rino and D Borges 2005 *J. Phys. Condens. Matter* **17** 5771

Journal of Physics: Conference Series **466** (2013) 012008 doi:10.1088/1742-6596/466/1/012008

Performance of air plasma spraying parameters in the fabrication of ZrO_2–10%Y2O_3–18% TiO_2 thermal barrier coatings

S Liscano[1] **and L Gil**[1]
[1] Departamento de Ingeniería Metalúrgica, Universidad Nacional Experimental Politécnica "Antonio José de Sucre", Puerto Ordaz, Venezuela

E-mail: sliscano@unexpo.edu.ve

Abstract. In this research, ZrO2–10%Y2O3–18% TiO2 thermal barrier coatings (TBCs) were sprayed by employing an air plasma thermal spray (APS) technique following a 23 factorial design experiment in order to obtain the best favorable spraying conditions able to enhance their properties. The effect of the spraying parameters on the porosity, microhardness, microstructure and morphology were determined by using different techniques such as optical microscopy (OP), image analysis, Vickers indentation and scanning electron microscopy (SEM) technique, this latter coupled with X-Ray microanalysis (EDS). It was found that both the arc voltage and the powder feed rate, as well as their interaction had a significant effect on the values of the reported hardness and these results were related to the existing level of porosity in the coatings. It was concluded that the best coatings properties, for the level of the variables studied in this work, could be obtained if the arc voltage is maintained at 36 V, the arc current at 900 A and the powder feed rate to 24 g/min.

1. Introduction

Plasma-sprayed yttria-stabilized zirconia (YSZ) continues to play an important role in enhancing performance of components for different industrial applications [1, 2]. Tailoring the microstructure and properties of these thermal barrier coatings towards achieving both prime reliance and manufacturing reproducibility is a complex task due to the multitude of interrelated parameters that influence both the plasma spray process and the deposit formation dynamics [3, 4]. The atmospheric plasma sprayed process involves melting of the feedstock materials in a plasma plume and rapidly transporting these molten particles to the substrate, where rapid solidification of individual particles occurs upon impingement. The successive build-up of these "splats" results in a layered arrangement in the coating, analogous to a brick-wall-like structure where the splats are intertwined in complex arrays [3]. This splat-based layered microstructure leads to an intrinsic anisotropy of the coating in the direction perpendicular to the spray direction. The splats are separated by interlamellar pores resulting from rapid solidification of the lamellae, globular pores formed by incomplete inter-splat contact or around unmelted particles as well as intrasplat cracks, which are due to thermal stresses and tensile quenching stress relaxation. The proper characteristic of the TBC originates from the existing porosity, the microcracks and their inherent microhardness [4, 5]. Although numerous studies involving processing parameters effects on the characteristics of such deposits have been carried out, a rigorous analysis on the effects of the APS parameters on the morphology and the mechanical properties of

YSZ plasma sprayed coatings employing the Experimental Design study has been little attempted; particularly, for the chemical composition of the powders used in the present investigation, which implies the addition of TiO2 to the traditional ZrO2-YO1.5 system, addition which is related to the enhancement of toughness, phase stability and sintering resistance, all of which are desirable for TBC applications [7].

The present study was conducted with the aim of the relationship between different deposition parameters of APS ZrO2–10%Y2O3–18% TiO2 coatings and their microstructural features and microhardness. The empirical relationships amongst the independent variables were achieved by employing both, statistical experimental design methods and the response surface methodology, described in detail elsewhere [7].

2. Experimental details

The TBCs, composed of a bond coat and a top coating, were industrially deposited by air plasma spray (APS) onto AISI 1020 steel coupons of 19.05 mm diameter. The powder used for the bond coating was Ni-5%Mo-5.5%Al (Metco® 447NS), which was deposited by using a Sulzer Metco® model 5P-II torch (manual operation). A Praxair model SG-100 torch coupled to mechanical arm was used to deposited ZrO2–10%Y2O3–18%TiO2 (Metco® 143) with powder size -75 +5 μm, as top coating. The industrial operational parameters are presented in table 1.

In this research, a 2^3 factorial design experiment was employed to establish the effects of the variables on the coating quality. Three operational parameters were varied: the arc current, the arc voltage and the powder feed rate, and setting to two levels and employing a central point (obtained of conditions of plant), which implies that nine experiments, were necessary for to explore the variation of these variables at the chosen levels. Two replicate experiments were carried out in each case. The test matrix is displayed in table 1.

Table 1. Industrial operational parameters and Test matrix for the 23 factorial experiment design.

Run N°	Arc current (A)	Arc voltage (V)	Powder feed rate (g/min)	Spray Distance (mm)	Ar Flow Primary (SCFH)	H2 Flow Secondary (SCFH)
Conditions of plant (Central point)	850	35,5	20	76	75	15
1	800	35	24			
2	900	35	24			
3	900	36	24			
4	800	36	24			
5	800	35	16			
6	900	35	16			
7	900	36	16			
8	800	36	16			

The experiments were carried out following a random sequence and the data were obtained for the response properties as: porosity and microhardness. These results were subject to an analysis of variance (ANOVA), based on F-test, in order to measure whether a factor contributes significantly to the variance of a response. The degree of the influence was represented as a frequency histogram. Response surfaces methodology (RSM) was used to describe empirical relationships amongst the three independent variables investigated. The statistical analysis of the result was conducted by using a commercial statistical analysis package [7].

In order to have reproducible results for the metallographic apparent porosity studies, samples cross section have been prepared following the procedure recommended in the literature for the thermal spray coatings [9]. The porosity was determined by carrying out the study of twenty fields captured on the cross section of the sample by employing digital image analysis (IA32 LECO Image Analysis System, USA). The microstructure of the coatings was examined by using a scanning electron microscope (Quanta FEG 600 SEM) coupled with a standard energy dispersive X-ray microanalysis attachment (EDAX). The hardness of the coatings was obtained using a Vickers indenter with the load of 300g applied for 15 seconds on the sample cross section. Results of ten measurements were averaged to obtain the reported microhardness values.

3. Results and discussion

3.1. Porosity

The average porosity and standard deviations values, as revealed by image analysis, are listed in Table II. As it can be observed, the average porosity of these coatings varied ranged between 14.7% to 26.5 %. From the analysis of the variance (ANOVA) it was found that arc current (A) and powder feed rate (C) were the only independent variables that contributes significantly to the variance of the response, at 95 % confidence level, for the interval studied in the present investigation. This could be observed in figure1 (a), where the frequency histogram shows the influence of each factor on the response function, i.e. on porosity.

Table 2. Summary of the characteristic results of the APS coatings for the 23 factorial experiment design.

Run N°	Microhardness ($HV_{0,3}$)	Porosity (%)
1	330	17,8±2,1
2	486	17,3±3,2
3	798	14,7±2,5
4	734	26,5±3,7
5	895	26,3±0,8
6	788	16,5±0,6
7	775	22,3±3,4
8	828	24,5±2,4
(central point)	591	18,5±3,2

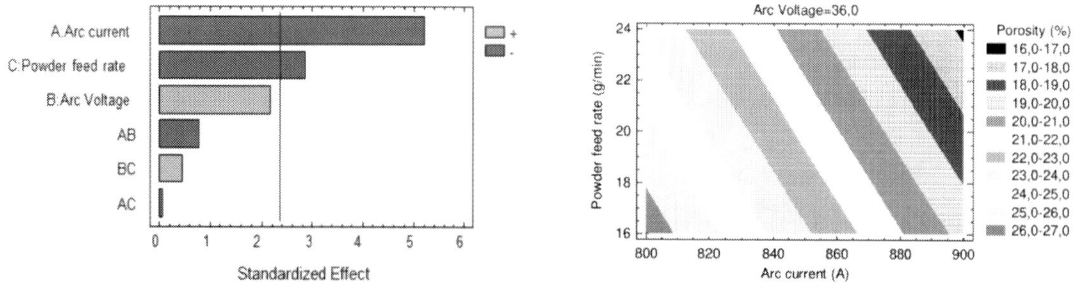

(a) **(b)**

Figure 1. (a) Frequency Histogram shows the Influence of each factor on the response function, i.e. on porosity. (b) Response Surface Contours of porosity for a constant value of Arc Voltage= 36,0 V and like a funtion of variation of Powder feed rate (g/min) and Arc current (A).

Since no substantial differences in porosity were found when the arc voltage was varied between 35V to 36 V, figure. 1 (b) represents the typical contours of porosity corresponding to the APS ZrO2–

10%Y2O3–18% TiO2 coatings as a function of the both arc current and powder feed rate values for a constant value of the arc voltage of 36V. For example, the figure 1 (b) shows that an increase of the arc current from 800 to 900 A, produced a decrease in the average porosity of 30%, when powder feed rate was fixed in 24 g/min. This result is expected since an increase in arc current develops more heat available in the spraying jet. Measurements performed by Vardelle et al. [9] when spraying alumina powder have indicated that a variation of 10% in the value of the arc current leads to a variation of 20% in the value of the process thermal efficiency. On the other hand, in the same figure, when powder feed rate varies from 16 to 24 g/min produces a decrease in the average porosity of 16% when the arc current is fixed in 900 A. These results corroborates those reported by Fleury et al. [10], where the porosity level linearly decreases as the powder feed rate increases. As expected, excessive feed rates may result in inadequate or partial melting of the powder. However, the powder feed rate should be adjusted to avoid the over-oxidation of the powder.

3.2. Microhardness

The hardness values of the coatings obtained during different tests were listed in table 2. As it can be observed, the average hardness values corresponding to the nine processing conditions varied between 330 to 895 HV300. Similar hardness values were reported by Shin [5], for coatings of plasma-sprayed ZrO2 coatings. From the ANOVA test, it was found that hardness depends of the powder feed rate (C), the arc voltage (B) and the interactions between the arc voltage and the powder feed rate (BC) and between arc current and the powder feed rate (AC), since they were determined as being significant from the statistical point of view, as shown in Figure 2a. The results of empirical modeling of the microhardness are also presented in figure 2. This figure presents the contours of micohardness for two conditions each one with constant value of the powder feed rate: 16 g/min (figure 2 (b)) and 24 g/min (figure 2 (c)). The axes, in both cases, correspond to the arc voltage (horizontal axis) and arc current (vertical axis).

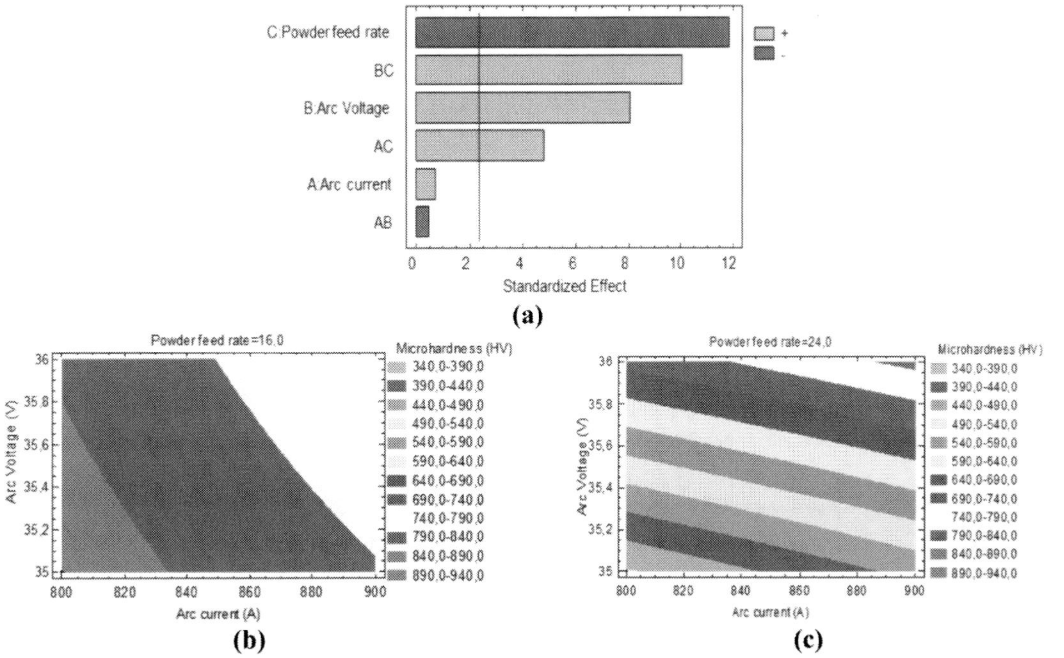

Figure 2. (a) Frequency histogram showing the influence of the variables and their interactions on microhardness. (b) Response Surface Contour of microhardness for the variables, current and voltage, the speed of the powder was maintained between values 16g/min. (c) 24g/min.

From the analysis carried out in figures 2 (b) – (c) it could be observed that power feed rate has a marked influence on the hardness value of the coating. In the figure 2 (b), at value constants of arc voltage and power feed rate in 16g/min, the hardness value starts to decrease when the value of the arc current is increasing. In contrast, when the powder feed rate increases to 24g/min, hardness behavior becomes opposite: increases with the rise of current arc (figure 2 (c)). This behavior in the mechanical properties has been attributed in the literature to interaction of other variables such as those related to the existing phases in the coating microstructure, as well as the adhesion between splats [10].

4. Conclusions

APS process can be investigated efficiently using methods based on factorial experimental designs. Empirical modeling could be successfully used to identify parameters that have a significant influence on coating properties. Analysis of the coatings characteristics has showed that the arc current, the arc voltage and the powder feed rate as well as their interaction had significant effects on the porosity and microhardness. In this study, the lowest porosity levels were obtained for the case of Run N°3 (highest parameters value), while Run N°5 (lowest parameters value) produced the hardest coating.

References

[1] Nusair Khan A and Lu J 2009 *J. Mat. Processing Technol.* **209** 2508
[2] Gómez J, Salazar A, Múnez C, Utrilla V and Poza P 2007 *Bol. Soc. Esp. Ceram.* **46** 232
[3] Kulkarni, Z Wang, T Nakamura, S Sampath, A Goland, H Herman, J Allen, J Ilavsky, G Long, J Frahm and RW Steinbrech 2003 *Acta Materialia* **51** 2457
[4] Mercer C, Williams J R, Clarke D R, Evans A G 2005*Acta Materialia* **53** 1029
[5] Shin J-H, Lim D-S, Ahn H-S 2000 *Surf. Coat. Technol.* **403** 133
[6] Schaedler T, Leckie R, Krämer S, Evans A, Levi C 2007 *J. of the American Ceramic Society* **90-12** 3896
[7] Statgraphics Plus version 5.1 2000 *Statistical Graphics Corporation* (Estates Unites)
[8] Blann G 1991 *Proc. of the Fourth National Thermal Spray Conference* (Estates Unites: ASM International)
[9] M Vardelle, A Vardelle, P Fauchais, K-I Li, B Dussoubs and N Themelis 2001 *J. Therm. Spray Technol.* **10-2** 267
[10] Fleury E, Lee S M, Kim W T, Kim D H 2000 *J. of Non-Crystalline Solids* **278** 194

Journal of Physics: Conference Series **466** (2013) 012009 doi:10.1088/1742-6596/466/1/012009

Aluminum-silicon co-deposition by FB-CVD on austenitic stainless steel AISI 316

J L Marulanda[1,2], F J Perez [2] and A Remolina-Millán[3]
[1] Grupo de Investigación en Materiales Avanzados -GIMAV, Faculty of Mechanical Engineering, Universidad Tecnológica de Pereira, Pereira, Colombia.
[2] Grupo de Investigación en Ingeniera de Superficies y Materiales Nano-estructurados, Faculty of Chemical Sciences, Universidad Complutense de Madrid, Madrid, España.
3 Grupo de Investigación – GIDETECHMA, Faculty of Mechanical Engineering, Universidad Pontificia Bolivariana seccional Bucaramanga, Floridablanca, Colombia.

E-mail: jlmarulanda@utp.edu.co

Abstract. Aluminum-silicon coatings were deposited on stainless steel AISI 316 in the temperature range of 540 to 560°C by CVD-FBR. It was used a fluidized bed with 2.5% silicon and 7.5% aluminum powder and 90% inert (alumina). This bed was fluidized with Ar and as an activator a mixture of HCl/H2 in ratios of 1/10 to 1/16. Furthermore, the deposition time of the coatings was varied between 45 minutes to 1.5 hours, with a 50% active gas, neutral gases 50%. Thermodynamic simulation was conducted with the Thermocalc software to get the possible compositions and amount of material deposited for the chosen conditions. The coatings presented the follow compounds FeAl2Si, FeAl2 and Fe2Al5. Aluminum-silicon coatings were heat treated to improve its mechanical properties and its behavior against oxidation for the inter diffusion of the alloying elements. The heat treatment causes the aluminum diffuse into the substrate and the iron diffuse into coating surface. This leads to the transformation of the above compounds in FeAl, Al2FeSi, Cr3Si, AlFeNi and AlCrFe.

1. Introduction

The Chemical Vapour Deposition (CVD) involves the chemical reaction of a precursor gas mixture inside a vacuum chamber for depositing a thin layer on the substrate. The reaction byproducts are evacuated outwards [1, 2]. This technique uses the chemical reactions of the precursor gases which are activated by heating, plasma, electromagnetic radiation, etc. The chemical deposition reactions can be classified into homogeneous and heterogeneous. Chemical vapor deposition is a very versatile process which allows several applications. Some applications are for protecting materials against wear, corrosion and oxidation at high temperature. It also allows the deposition of amorphous elements, mono crystalline, polycrystalline thin layers, dielectrics, for microelectronics and optoelectronics in energy conversion devices. Also can be manufactured ceramic fibers and composites ceramic matrix, as well as nanostructured materials for the electronics industry and biotechnology [3-5].

The chemical vapor deposition fluidized bed is a variant of the technique of chemical vapor deposition that combines the advantages of thermal activation by heating and the fluidized bed. The CVD-FBR takes advantage of the fluidized beds, as are the high transfer heat and mass between the gas, the bed and samples immersed in the reactor. The CVD-FBR allows a more uniform temperature and good mixing of reactive gases with the fluidized particles. This produce a fast reaction of the activated species in all the bed, because there fluidization in excellent contact between the solid particles and the gaseous fluidizing medium. This helps reduce the temperature and operating time compared to other processes such as pack cementation. The deposition reaction between the chemical components of the donor and the activator precursor vapors form the element layer, the layer being formed by the reaction of the substrate with the vapors produced from the precursor [6-8].

2. Results

Pre-eliminate tests were conducted in the CVD-FBR reactor using a fluidized bed formed by a 5% aluminum powder, 5% silicon powder and 90% bed inert (alumina). It was fluidize with 61% argon and as gas activators a mixture of hydrogen 37.1% and hydrochloric acid 1.9%, for a time of 90 minutes at 580°C. This was done to find the best conditions of the deposition process on the stainless steels. The above conditions are producing good coatings with a thickness approximately of 10 microns (±4 µm). The amount of silicon is incorporated into the coating is below 0.5%. Because the amount of silicon was minimal in the co-deposition aluminum-silicon we analyzed the simulation thermodynamics and literature results [5-10] and we determinate the lowest the deposition temperature was 540°C. This is due to the greater amount of ClH_3Si present in the process of deposition. Therefore there is a greater possibility the silicon is deposited and best thermal activation increasing the probability of obtain a good aluminum-silicon coating. The best aluminum-silicon coatings were obtained when deposition was carried out at 540°C with a ratio of active gases HCl/H_2:1/15,3 and 2.5 g of aluminum and 7.5 g of silicon in the bed and a 50% active gases, neutral gases 50%. The best samples are reached in the early stages of fluidization.

Figure 1. Morphological surface (a) and cross-sectional (b) image of aluminum-silicon coating.

The figure 1. Shows the SEM results for aluminum-silicon coating. The fig 1a shows a cone-shaped morphology of the surface. The chemical composition of the coating is 54.59% Al, 24.35% Fe, 8.05% Ni, 7.67% Cr, 3.62% Si, 0.62%Mg and 1.10% Mo. The aluminum is found like $FeAl_2Si$, Fe_2Al_5 and $FeAl_2$. These coatings were heat treated to improve its mechanical properties by the interdiffusion of the alloying elements. These coating heat treated are formed by compounds of FeAl, Al_2FeSi, Cr_3Si, $AlCrFe_2$, and AlFeNi. The FeAl is the greater extent on the outside of the layer. The coatings no heat treatment have a higher amount of aluminum than heat treated coatings. After heat treatment disappear the compounds of Al_5Fe_2, FeAl, decreasing the percentage of aluminum and increasing the percentages of iron, chromium and nickel.

The heat treatment favoring the transformation of aluminum rich phases, in other more iron content, thereby improving the coating mechanical properties and corrosion resistance. Finally, it can be concluded that by the process FBCVD can obtain coatings of aluminum-silicon 10 microns thick, for the deposition temperature of 540 °C and a deposition time of 1 hour. After a heat treatment such layers can grow up to 14 um thick, because appears a inter diffusion zone within the substrate. The thermodynamic simulation is a good tool to select the working conditions and know the aluminum and silicon halides that are taking the deposition of the coating.

3. Conclusions

Were optimized deposition conditions of the coatings of Al-Si by FBCVD at temperatures below 600 °C. These temperatures do not affect the microstructure of the steels. We obtained thicknesses greater than 14 microns. The Temperature and flow of the precursor gases are directly related to the kinetics of growth of the coatings. The temperature of deposition of the coatings should not exceed 560 °C because they begin to chlorinate the coatings and not able to obtain good layers. This paper had the highest coating thickness for the aluminum-silicon alloy at 540 °C to 60 minutes and with active gas mixture of 50% argon, 46.83% hydrogen and 3.17% HCl, with a bed of 2.5 g Al, 7.5 g Si, and 90g of Al_2O_3. conductivity. Any additional amount of HEM produces a negative effect on PANI/HEM conductivity.

References

[1] K Oura, V G Lifshits, A A Saranin, A V Zotov and M Katayama 2003 *Surface Science an Introduction* (New York: Springer-Verlag Berlin Heidelberg)
[2] J Marulanda, F Pérez, S Castañeda 2011 *Discussion meeting on the development of innovative iron aluminium alloys* (España: Lanzarot)
[3] C Kleijn, R Dorsman, K Kuijlaars, M Okkerse, H van Santen 2007 *Journal of Crystal Growth* **303** 362
[4] S Castañeda, F Bolívar, F Pérez 2010 *Oxidation of metal* **74** 61
[5] L Sánchez, F Bolívar, M Hierro, J Trilleros, F Pérez 2007 *Surface and Coatings Technology* **201** 7626
[6] S Castañeda, J Marulanda, F Pérez 2011 *Discussion meeting on the development of innovative iron aluminium alloys* (España: Lanzarot)
[7] N Bahlawane 2001 *Thin Solid Films* **394** 298
[8] F Pérez, F Pedraza, M Hierro, P Hou 2000 *Surface and Coatings Technology* **133** 338
[9] F Bolıvar, F Pérez, M Hierro, J Trilleros, L Sánchez 2007 *Scientia et Technica* **36** 619
[10] F Pérez, S Castañeda 2007 *Surface and Coatings Technology* **201** 6239

Journal of Physics: Conference Series **466** (2013) 012010 doi:10.1088/1742-6596/466/1/012010

Study of the structural and electronic properties of YC using DFT: The true ground state is a NiAs-like structure

G P Abdel-Rahim[1,2], J A Rodríguez M[2] and M G Moreno-Armenta[3]
[1] Grupo PERSEO, Universidad Antonio Nariño Sede Bogotá, Colombia.
[2] Grupo de Estudio de Materiales - GEMa, Departamento de Física, Universidad Nacional de Colombia, Sede Bogotá, Colombia.
[3] Centro de Nanociencias y Nanotecnología, Universidad Nacional Autónoma de México, Ensenada, BC, México.

E-mail: pabdelrahim@gmail.com

Abstract.. We study the structural and electronic properties of YC in volume using density functional theory (DFT) within the generalized gradient approximation (GGA), using the scheme of Wu-Cohen 2006 and Tran et al. 2007. Several crystal structures were considered: Nickel Arsenide (NiAs), Sodium Chloride (NaCl), Cesium Chloride (CsCl), and zincblende (ZB). A new fact that we present in this paper is the inclusion of the NiAs-like structure, which is the true ground state (GS) for YC. We calculated the density of states (DOS) and the band structure and found that YC is non-magnetic and its behavior is metallic-like. The lattice parameter alatt is 3.69 Å and the c/a = 1.47. Cohesion energy (Ecoh) is -12.90 eV, which is very close to Ecoh of the NaCl structure. Therefore, YC exists in these two structures. Our results with respect to alatt, bulk modulus (B), Ecoh, and the main features of the electronic properties are in good agreement with those found by other researchers. Other researchers found a transition from NaCl to CsCl, but we found a new transition from NiAs to CsCl, where the volume diminishes ~10% and its transition pressure (PT ~79 GPa) is very close to the 80 GPa of the former. The contraction can fracture the material if it is worked on near the transition. For pressures before and after the transition, YC maintains non-magnetic and metallic behaviors.

1. Introduction

The 4d-transition metal carbides (4d-TMC) are studied mainly due to their many industrial applications for mechanical parts and the fact that they are highly resistant materials. Additionally, 4d-TMC preserve various types of chemical bonds, such as the ionic, metallic and covalent ones. Other properties that increase the great number of industrial applications are their high degree of hardness, high melting point, and good electrical and thermal conductivity [1-5]. Pooja Soni [1] studied the structural properties in phases such as NaCl and CsCl using DFT with the exchange and correlation potential GGA of Perdew, Burke (GGA-PBE). They found a phase transition from NaCl to CsCl, at PT~80 GPa. The study by K. K. Korir [2] of the 4d-TMC and 4d-TM nitrides in phases NaCl, ZB, and wurtzite, using DFT with the Local Density Approximation (LDA) method and GGA-PBE, showed that the experimental values of Ecoh of the carbides or nitrides are close to those calculated with GGA-PBE. In Jemeson's study, the electronic, structural, and elastic properties of YC and ScC in the NaCl phase are shown, which were calculated using LDA and GGA-PBE96.

2. Method

Calculations were carried out using WIEN2k code [6], within the framework of DFT, with the GGA approximation of Perdew, Burke, and Ernzerhof [7]. Because phases can exhibit a metallic behavior, a large number of k-points over the Irreducible Brillouin Zone (IBZ) were taken into account. We used about 135 k-points in IBZ. The muffin-tin radii were 1.99 and 1.77 a.u. Calculations were considered convergent if the difference of the total energy of successive iterations were consistent within a range of 10-4 Ry. The base functions inside the muffin-tin spheres were expanded up to lmáx = 10, on the basis of spherical harmonics, and the energy for the expansion of the wave function on the plane wave basis was limited by Rmt × Kmax = 7.0, where Rmt is the minimum muffin-tin radius. When it was necessary to optimize the c/a ratio and the volume, we used a recurrent process to optimize them, which included a minimizing of the forces in order to find the best atomic positions; the process was considered finished when the difference between the c/a and the volume of two successive calculations was less than 0.1 %.

3. Structural properties

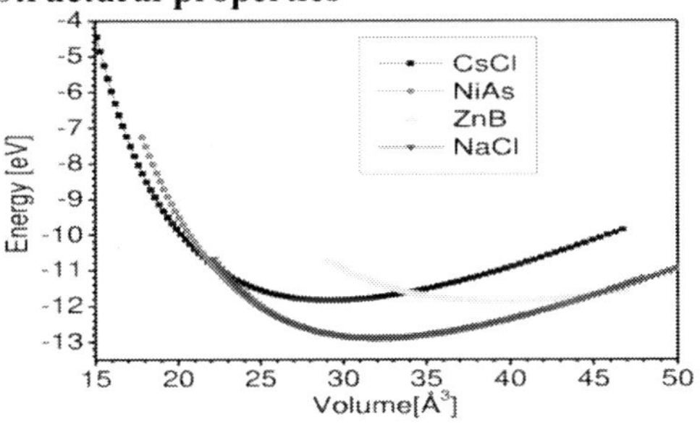

Figure 1. Energy vs. volume for the compound YC in several structures.

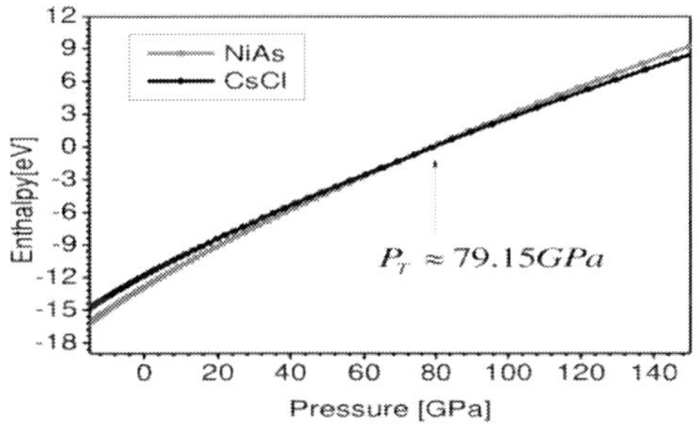

Figure 2. Enthalpy vs. pressure for NaCl and CsCl.

Table 1. Structural parameters of YC. B is the bulk modulus and B' its derivative with respect to pressure.

Structural Parameter	Structural Phase							
	NaCl		CsCl		ZB		NiAs	
		TW*		TW*		TW*		TW*
a_{min} [Å]	5.086[a],5.01[b],5.09[e] 5.11[c],5.19[d]	5.04	3.10[a]	3.08	5.47[b]	5.39		3.69
c/a	---	---	---	---	---	---		1.47
B [GPa]	124.18[a],124.3[b],1262[e]	122.40	120[a]	117.8	80.50[b]	88.41		119.7
B'	4.14[a], 4.451[e]	4.04	4.12[a]	3.50	11.53[b]	3.36		3.15
E_{coh} [eV]	12.03[b],12.79[d]	-12.87		-11.84		-11.88		-12.90

[a] Reference 1, [b] Reference 2, [c] Reference 4 (experimental), [d] Reference 5, [e] Reference 3, * TW meaning this work.

Figure 1 shows the $E_{coh} = E_{comp} - E_C - E_M$ as a function of volume by unit formula. E_{comp} is the total energy by unit formula, and $E_{C/M}$ the total energy of the carbon/metal free atoms. Other authors use $-E_{coh}$ instead of the above definition. The solid lines are the adjustment to Murnaghan's equation state [8]. One can conclude, from figure 1, that the GS is in the NiAs-like phase. The structural parameters are shown in Table 1, and a comparison with the results of other researchers can also be observed. The E_{coh} in the structure NaCl is only 0.03 eV higher than for the NiAs-like structure. That difference is very small, and permits the material to exist simultaneously in the NaCl and NiAs phases. By contrast, the E_{coh} of the structure ZB is ~1 eV higher than the value of NiAs-like phase. Therefore, the ZB phase is almost impossible to find spontaneously. A second conclusion from Figure 1 is that according to the behavior of the curves of the NiAs-like phase and that of the CsCl-like phase, there is a high-pressure phase transition. Figure 2 shows that the pressure at which the transition occurs is PT ~ 79 GPa. Based on table 1, one additional conclusion about the structural properties of YC is that the a_{latt}, of the NaCl phase calculated with GGA-WC is under-estimated when compared with the results that one can find using GGA-PBE. Also, this fact is observed by means of comparison of the a_{latt} obtained by theoretical methods [1-3] and experimental ones [4] with those obtained using GGA-WC.

4. Electronic properties

This section is devoted to the study of the electronic properties of YC in the GS. Figures 3 and 4 show the total and partial DOS at pressure P=0, i.e. at the minimum of the function E vs. V in Figure 1. The energy is referenced taking as the origin the Fermi level (EF). Although the calculations included spin, we only show only one spin component, because both DOS were identical. Therefore, YC is non-magnetic in its GS. In these figures, we see three regions: the first is a narrow peak below -6.4 eV due mainly to the s-C orbital (see figure 3), plus a small contribution by the p-Y orbitals (see figure 4). The second region is located between EF and E~-4 eV, where we can see the upper valence band (VB), which is composed of 2p-like C orbitals (pz and px+py orbitals of C) and 3p-like Y and 3d-like Y orbitals. These orbitals are hybridized in this energy region. At EF, the DOS very quickly descends toward almost zero; this fact permits one to conclude that the material exhibits a metallic-like behavior. The last region is the conduction band, (CB), which is linked to the valence band (VB) without an energy gap, and this region is formed mainly by d-like Y orbitals. Figure 5 shows the bands at the GS and the metallic-like behavior is evident. The bottom VB (~ -7 eV to ~ -10 eV) is composed of two s-like C bands, and additionally there is a small component of a 3p-like Y orbital. There are two bands, because the NiAs phase has two lattice points in the Bravais lattice. Therefore, in figure 3, two small peaks are present in the DOS in this region. In this zone, the electron states show little dispersion, and they are localized in a narrow band of ~ 3 eV. In contrast, the upper VB and the bottom CB show great dispersion, which produces delocalized states.

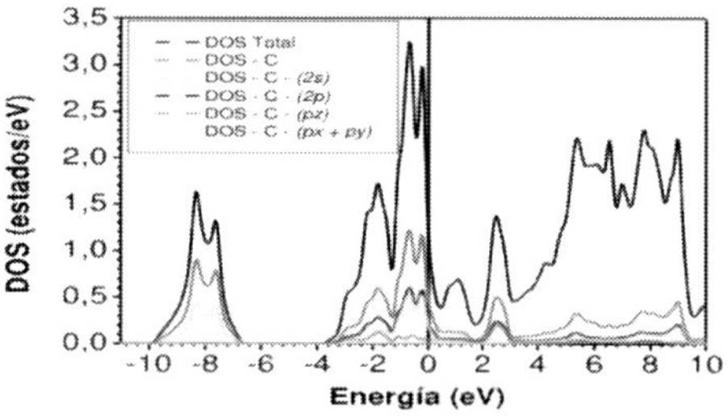

Figure 3. GS total and partial DOS due to C.

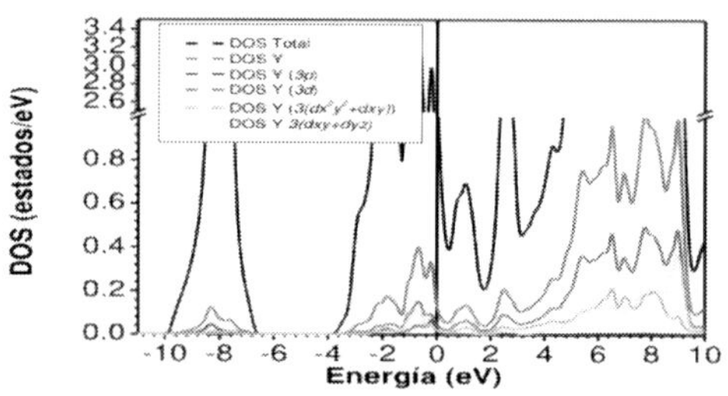

Figure 4. GS total and partial DOS due to Y.

5. The features of the high pressure transition

In this section, we analyze the structural and electronic changes before and after the transition between the NiAs and CsCl phases. Understanding properties in different phases is important because in applications YC can suffer a phase transition because of increasing pressure or temperature. Changes in the electronic properties are determined by changes in the length of the bonds, the hybridizations, and the symmetries. As we mentioned above, the transition occurs at PT~79 GPa. There is a volume reduction of ~10% from 23.66 Å3 to 21.23 Å3, and this compression can fracture the crystal; this is not desirable. Therefore, the material must be worked on at a pressure below 79 GPa. Figures 6 and 7 show the bands before and after the transition, respectively. In both phases it is non-magnetic and metallic. The bottom VB changes from two bands to only one. This modification is due to the different number of lattice points in the Bravais lattices. In the upper VB and in the CB, the hybridizations are similar to the hybridizations in the GS, but other changes of the bands are great and can be observed in the figures.

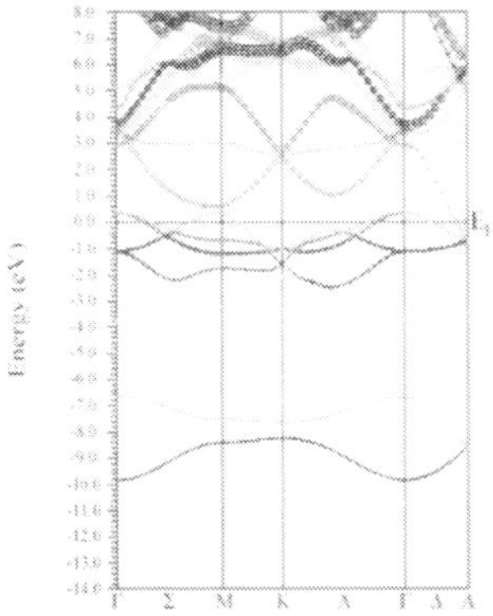

Figure 5. Band structure in the phase NiAs at P = 0 or GS.

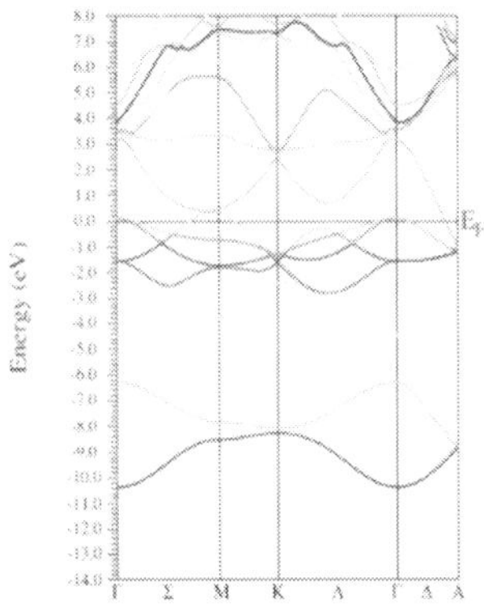

Figure 6. Band structure before of the transition. The structure is NiAs at pressure near but below 79GPa.

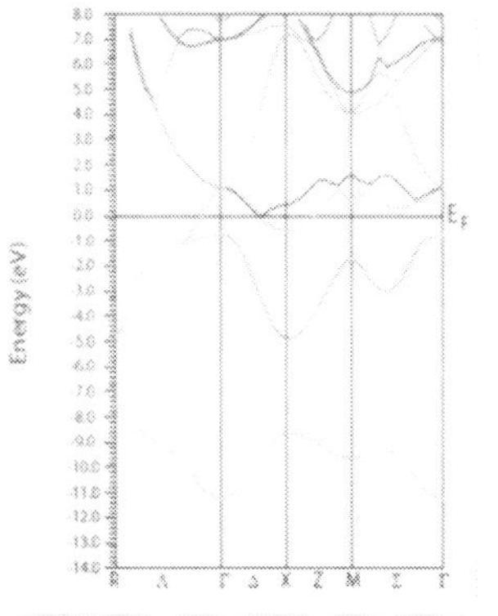

Figure 7. Band structure after the transition. The structure is CsCl at high pressure near but over 79GPa.

6. Conclusions
Calculations of structural and electronic properties of YC were carried out by means of DFT using the Wu-Cohen GGA approximation. Several phases were considered, and the GS is in the NiAs-like phase. YC, in the GS, is non-magnetic and metallic. alatt = 3.69 Å for this phase and the c/a = 1.47. Its cohesion energy is about -12.90 eV. It is always possible that the YC exist simultaneously in two structures: NaCl and NiAs. A new phase transition (at PT~79 GPa) from NiAs to CsCl was found, where the volume reduces by about 10%; the pressure transition is very close to 80 GPa, which is the pressure for the change from NaCl to CsCl. During the transition, YC could fracture if it is worked on near transition. Before and after the transition, YC is non-magnetic and metallic.

Acknowledgments
Financial support was given by DIB, Fac. de Ciencias, U. Nacional, and U. Antonio Nariño. Calculations were performed in the DGCTIC-UNAM Supercomputing Center in Mexico and Cluster of Grupo GEMa, Dept. de Física, U. Nacional.

References
[1] Pooja Soni, Gitanjali Pagare, Sankar P Sanyal 2011 *Journal of physics and Chemistry of solid* **72** 810-16
[2] K K Korir, G O Amolo, N W Makau and D P Joubert 2011 *Diamond Related Materials* **20-2** 157-64
[3] Jameson Maibam, B Indrajit, Sharma, Ramendu Bhattacharjee, R K Thapa, R K Brojen Singh 2011 *Physica B* **406** 4041-45
[4] P V Villars, L D Calvert 1985 *Pearson`s Handbook of Crystallographic Data for intermetallic Phases* (Ohio: ASM International, Materials Park)
[5] A F Guillermet, J Haglund, G Grimvall, M Smith 1192 *Phys. Rev.* B **45** 11557
[6] P Blaha, K Schwarz and J Luitz 1999 *A Full Potential Linearized Augmented Plane Wave Package for Calculating Crystal Properties* (Austria: Techn Universität Wien)
[7] J P Perdew, K Burke, M Emzerhof 1196 *Phys. Rev. Lett.***77** 3865
[8] F D Murnaghan 1944 *Proc. Natl. Acad. Sci.* **30** 5390

Journal of Physics: Conference Series **466** (2013) 012011 doi:10.1088/1742-6596/466/1/012011

Synthesis of hybrid sol-gel coatings for corrosion protection of we54-ae magnesium alloy

C A Hernández-Barrios[1], N Z Duarte[2], L M Hernández[2], D Y Peña[1], A E Coy[1] and F Viejo[2]
[1] Escuela de Ing. Metalúrgica y Ciencia Materiales, Universidad Industrial de Santander, Bucaramanga, Colombia.
[2] Escuela de Ingeniería Química, Universidad Industrial de Santander, Bucaramanga, Colombia.

E-mail: carloshernandezbarrios@hotmail.com

Abstract. The present work shows some preliminary results related to the synthesis, characterization and corrosion evaluation of different hybrid sol-gel coatings applied on the WE54-AE magnesium alloy attending to the two experimental variables, i.e. the precursors ratio and the aging time, which may affect the quality and the electrochemical properties of the coatings resultant. The experimental results confirmed that, under some specific experimental conditions, it was possible to obtain homogeneous and uniform, porous coatings with good corrosion resistance that also permit to accommodate corrosion inhibitors.

1. Introduction

Nowadays, it is well known that magnesium alloys are excellent candidates to reduce weight of the manufactured component without sacrificing its mechanic resistance. For this reason, these alloys are used in the transport industry, for the construction of train and aircraft fuselages, among others components for sport vehicles. One of the most important magnesium alloys is the WE54 alloy (Mg-Y-RE) that finds application at relatively high temperatures (up to 300ºC) for the construction of power systems, transmissions, and high performance cars due to their high strength and good creep resistance [1, 2].

Nevertheless, magnesium, as element, exhibit low chemical stability with a standard electrode potential of -2.37V, it being the most active metallic element utilized for structural applications [3-5]. This fact finally results in a severe decrease in corrosion resistance of their alloys in most of the frequent environments where magnesium might find potential application.

Among the protective coatings available for corrosion protection of magnesium alloys, particular interest has been focused on the sol-gel route, which is characterized by the formation of ceramic films with excellent corrosion resistance through successive reactions of hydrolysis and condensation of the corresponding precursors [6, 7]. With regard to other conventional methods, coatings prepared through this technique offer great advantages, i.e. great adhesion, low surface preparation and processing temperatures, use of relatively simple and inexpensive equipment, the possibility of application on complex shaped surfaces, etc.

In this regard, different precursors have been used to synthesize inorganic films such as SiO_2, TiO_2 or ZrO_2. However, due to their ceramic nature, these films cannot be growth without cracking for more than 1 micron. Recent studies are investigating the possibility of producing hybrid coatings, consisting in the addition of different organic precursors to the inorganic sol, which present organic groups such as epoxy (GPTMS) or amino (APTES), etc. that promotes crosslinking during of the sol-gel network and reduces both temperature and time during the curing process. As a consequence, the sol-gel films exhibit lower cracking susceptibility, high density and flexibility, and provide excellent mechanical properties and corrosion resistance [8, 9]. Further, hybrid sol-gel coatings are also interesting because of their porous structure since it permits to accommodate different kind of additives, e.g. corrosion inhibitors, such as rare earths elements, which provides an active self-curing effect [10-12].

Based on these statements, the aim of the work was to synthesize, characterize and evaluate the corrosion resistance of different hybrid sol-gel coatings applied on WE54-AE magnesium alloy.

2. Experimental

The material used in the present investigation was the WE54-AE magnesium alloy with nominal composition Nd: 1.5-2.0%; Y: 4.8-5.5%; RE: 1.0-2.0%; Zr: 0.4% y Mg: balance (%wt). Hybrid sols were prepared from a mixture of the inorganic precursor tetraethylorthosilicate (TEOS) and the organics precursors glycidoxypropyltriethoxysilane (GPTMS) and 3-aminopropyltriethoxysilane (APTES). Two experimental variables were fixed: the molar relation APTES:GPTMS (from 0.00 to 0.25:1) and the aging time (from 15 minutes to 24 hours). Later, sol-gel coatings were obtained by deposition of these sols following the dip-coating route. Finally, coatings were cured in two steps where temperature was increased progressively to prevent cracking: 60ºC (4 hours) and 120ºC (2 hours). Coatings were characterized using Scanning Electron Microscopy (SEM). Finally, corrosion resistance of the sol-gel coatings was evaluated by anodic potentiodynamic polarization test in aerated 0.1M NaCl solution after 30 min of stabilization [13].

3. Results and discussion

Figure 1 shows the electron micrographs of the hybrid sol-gel coatings performed on the WE54-AE magnesium alloy for different APTES: GPTMS molar ratios and aging times.

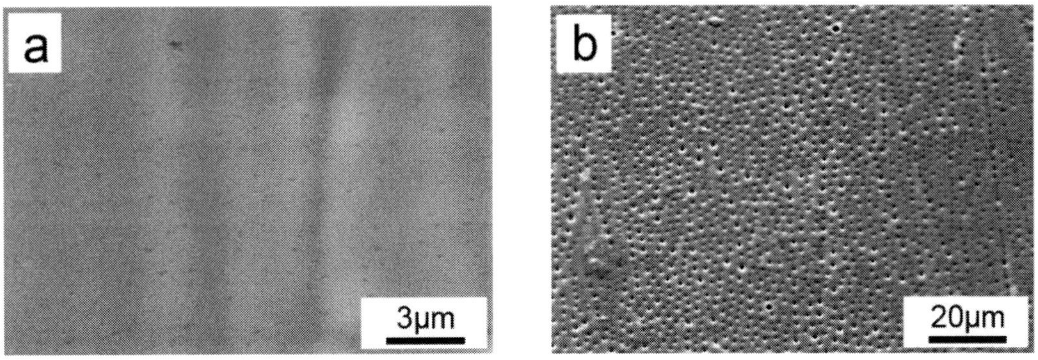

Figure 1. Electron micrographs of hybrid sol-gel coatings with APTES:GPTMS molar ratio and aging time of: (a) 0.00:1-24h and (b) 0.25:1-6h.

An increase of the aging time promoted the formation of continuous and uniform films, figure 1 (a). On the other hand, the addition of the APTES precursor accelerated the gelation process, which allowed obtaining coatings also uniform but with high grade of porosity, figure 1 (b). The latter was attributed to the presence of the NH_2 groups in the APTES precursor that shifts pH toward higher values and favoured the condensation process. This fact prevents the creation of an organized sol-gel structure and the continuous evaporation of the solvent (ethanol), which is finally trapped within the network. After dip-coating and during the curing process, the solvent rapidly evaporated resulting in the consequent porous structure.

Figure 2 shows anodic potentiodynamic polarization curves of the parent alloy and hybrid sol-gel coatings with different APTES: GPTMS molar ratios and aging times.

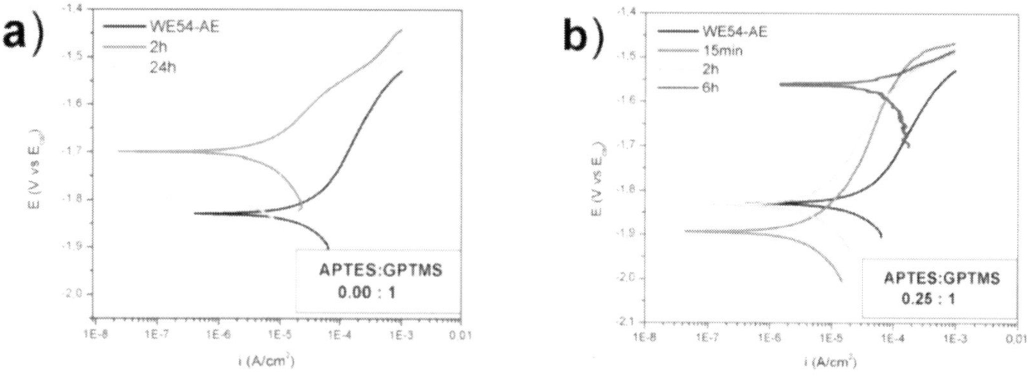

Figure 2. Anodic potentiodynamic polarization curves of the parent alloy and hybrid sol-gel coatings with APTES:GPTMS molar ratio of: (a) 0.00:1 and (b) 0.25:1 for different aging times.

In general, the hybrid sol-gel coatings enhanced the corrosion resistance of the parent alloy regardless of the addition of the APTES precursor, diminishing the corrosion current density by about one order of magnitude (See figure 2 (a)). Further, the addition of APTES, at short aging times, resulted in greater corrosion performance with also wider passivation range (Epit-Ecorr) (see figure 2 (b)). However, for longer aging times, the great porosity of the sol-gel network permits the easy diffusion of the corrosive species that promptly reaches the surface of the bulk alloy.

4. Conclusions

The sol-gel synthesis using mixture of inorganic-organic precursors allowed obtaining coatings with excellent surface characteristics of continuity and homogeneity. As a result, the corrosion performance of the WE54-AE magnesium alloy was significantly improved in terms of corrosion current density and passive features. The aging time and the APTES: GPTMS molar ratio affect the morphological features and the electrochemical properties of the coatings obtained: an increase of the aging time resulted in more continuous and homogenous coatings, while the addition of APTES promoted uniform coatings but with great porosity and lower corrosion resistance.

Acknowledgements

The authors wish to thank to the Vicerrectoría de Investigación y Extensión of the Universidad Industrial de Santander, Colombia (Strategic Area of Materials, Code 5450) for the financial and technical support of the present work.

References

[1] Mike D 2001 *Environmental Effects on Engineered Materials* (Nueva York: Marcel Dekker Inc.)
[2] Horst F, Barry L 2006 *Magnesium Technology* (Germany: Springer)
[3] Ling G, Atrens A 1999 *Adv. Eng. Mater.* **1** 11
[4] Kulekci M 2008 *J. Adv. Manufacturing Technol* **39** 851-65
[5] Tamar Y and Mandler D 2008 *Electrochim. Acta.* **53** 5118-27
[6] Liu D et al 2001 *Biomaterials* **22** 1721-30
[7] Wang D and Bierwagen G 2009 *Prog. Org. Coat.* **64** 327-38
[8] Guglielmi M 1997 *J. Sol-Gel Sci. Technol.* **8** 443-9
[9] Zheng S et al 2010 *J Sol Gel Technol.* **54** 174–87
[10] Yue Xu et al 2007 *J. Rare Earths.* **25** 193-6
[11] Schem M et al 2009 *Corros. Sci.* **51** 2304-15
[12] Gigandet M et al 2008 *Surf. Coat. Tech.* **202** 2052-8
[13] ASTM G3-89 1999 *Standard Practice for Conventions Applicable to Electrochemical Measurements in Corrosion Testing* (Estates Unites: American Society of Testing and Materials - ASTM)

Journal of Physics: Conference Series **466** (2013) 012012 doi:10.1088/1742-6596/466/1/012012

Development of cellulose-polypyrrole microfiber membranes and assessment of their capability on water softening

C Barrera[1], A Arrieta[1], N Escobar[2], P Gañan[2] and C Castro[2]
[1] Grupo de Investigación de Desarrollo y Aplicación de Nuevos Materiales – DANM, Universidad Pontificia Bolivariana, Montería, Colombia.
[2] Universidad Pontificia Bolivariana, Medellín, Colombia.

E-mail: alvaro.arrieta@upb.edu.co

Abstract. The application of conducting polymer composite for water softening is based on the use of pyrrole`s electrochemical properties joined with the flexibility and relatively high surface areas associated with cellulose fibers, to develop a new hybrid material that exhibits the inherent proprieties of both components. This hybrid would allow to promote an ion exchange reaction between the composite membrane and the hard water. The cellulose membranes obtained from banana plant agricultural waste (raquis), were uniform with individual and well separated fibers. The fibers were encapsulated by a continuous coating of polypyrrole by an in situ oxidative chemical polymerization. The amount of polypyrrole deposited on the fiber increased by increasing the monomer concentration, behavior that was identified through the observation of differences on the intensity of the light to dark color shift that coated the fibers after the polymerization. The ion removal capability of the membrane coted with the conducting polymer was tested using an experimental device, finding reductions on the conductivity for hard water within 23 to 66 μs/cm after 6 hours of the assay.

1. Introduction

Conducting polymers (CP) represent a novel generation of organic materials which possess electrical and optical properties similar to the metals and inorganic semiconductor, but with the properties related to conventional polymers such as: ease of synthesis, flexibility, chemical inertness, low density, low cost, processability, among others [1, 2]. Different studies in the last few years has been focused on the applications of conducting polymers in electronic, environmental, biologic, as well as other fields, due to the variety of properties these materials possess [3, 4]. The CP's property of interest in this research is the electrochemical property, which can be defined as the ion exchange CPs undergo when they are subjected to an electrical stimulus in an electrolytic cell [5]. This property makes these materials an excellent possibility to develop ion exchange membranes with important applications in water softening. The development of polypyrrole-cellulose membrane provides the opportunity to generate new hybrid materials that exhibit the properties of the combined components. In this case the flexibility, resistance and surface area offered by the cellulose fibers, strengthened with the electrochemical properties of polypyrrole [6, 7].

2. Materials and methods

The cellulose membranes used in the experiments were provided by the Pontificia Bolivariana University's research group GINUMA. Pyrrole and other reactants were purchased from Sigma-

Aldrich. Conducting Polypyrrole-cellulose composites were obtained through insitu oxidative polymerization using two different procedures. In the first procedure the pyrrole was added directly to the cellulose fibers and in the second one, a solution of pyrrole with the dopant ion was prepared, and then this solution was added on the fibers. A complete description of the methods, reagent and concentrations used in each method is shown in table 1.

Both procedures were done for membranes with cylindrical and disk shapes and repeated three times for each method and shape analyzed. After polymerization the composite membranes were thoroughly washed with a solution 1:1 methanol/water in order to remove the excess of monomer, dopant ion or oxidant agent.

Table 1. Procedures followed to obtain the Polypyrrole-cellulose composites membranes.

Method	First step of procedure	Time of contact with the fibers	Second step of procedure	Time of contact
Direct deposition	Same amount of Pyrrole 98% and a solution of sodium p-toluen-sulfanate 2M (dopant ion), was added on the fiber to completely cover them.	45 min. Under ultrasonic treatment	1 ml of aqueous solution of ammonium persulfate 2M (oxidant agent), was added.	24 h
Solution deposition	Solution of Pyrrole 1M with sodium p-toluen-sulfanate 2M (dopant ion), mixed by ultrasonic treatment, and then added on the fiber to completely covered them.	45 min	1 ml of aqueous solution of ammonium persulfate 2M (oxidant agent), was added.	24 h

2.1. *Characterization of the Polypyrrole-cellulose material*

Membrane surface topography and the coating level of the fibers were imaged using a JEOL JSM-6490L Scanning electron microscope, operating at 15 kV and 80 A. Analysis was then performed on the cross section area of the samples. The structural characterization of composite material was performed using a Nicolet 6700 FT-IR Spectrometer. Spectra were recorded at the range of 4000 to 400cm−1 by accumulating 32 scans.

2.2. *Ion removal by the polypyrrole-cellulose membrane*

For this experiment a factorial design with two levels and two factors was used. An experimental device proposed by the research groups DANM and GINUMA, shown in figure 1, was used to assess the composite membranes′ ion removal application. As a hard water sample, a solution of 500 ppm of magnesium chloride and 500 ppm of calcium chloride was prepared.

A conventional three electrode configuration was used. The composite membranes were used as work electrodes. These three electrodes were placed in the upper container of the device, connected to potentiostat-galvanostat 2263 PARSTAT through which was loaded with a potential current of -0.5V to the immersed membrane being tested. The decreased conductivity of the hard water was measured with a conductimeter 315i WTW. Three membranes from each method and shape used were tested for six hours and the average value was calculated as the mean value.

Figure 1. Experimental device used to assess ion removal by the composite membrane.

3. Results and discussion
The cellulose membrane fibers obtained from agricultural waste were successfully encapsulated by a continuous coating of polypyrrole. The natural color of cellulose membranes before the polymerization was white, and after the chemical in situ polymerization of pyrrole on the fiber, all the membranes turned black. This dark color is characteristic of polypyrrole and shows the polymer deposition on the cellulose fibers. Preliminary assays with different concentrations of monomer and dopant ion shows that the amount of polypyrrole deposited on the fiber increased with increasing these two reactant concentrations. However, at higher concentrations of the monomer than the concentrations of the dopant ion, the solution gets a certain grade of immiscibility, requiring harder agitation.

3.1. Characterization of the Polypyrrole-cellulose material
The results of comparing composite material obtained through the two procedures used (direct deposition and solution deposition); indicate that the use of the second procedure leads to a more uniform coating of the fibers as is shown in figure 2.

The images in figure 2 (b) and (d) show the enlargement of the adherence of the polypyrrole to the cellulose fibers. As can be seen in figure 2 (b) the direct deposition of the monomer and dopant ion on the cellulose membrane promotes agglomeration of the polymer. This result may be due to better integration of the initial reagent that can be achieved through the solution deposition procedure. Regarding the two different shapes of membranes analyzed, there was no variation found on the polymerization behavior between the cylindrical and disk shapes. The tendency to form agglomeration of the polymer for direct deposition was observed in both shapes.

The spectra obtain for the polypyrrole-cellulose composite membranes show characteristic broad band of cellulose O–H group around (3375-3210) cm^{-1}, which confirm the presence of cellulose in the composite materials analyzed. The typical spectral features of the polypyrrole are also reported in the literature [2, 8]. There are absorption bands around (1644-1415) cm^{-1} for all the membranes tested; these bands are attributed to the C-C vibration in the pyrrole ring. The absorption bands, around (11121-1052) cm^{-1}, are characteristic of C-O functional groups. The differences of the wavelength of the bands in spectra indicate the different polypyrrole concentration obtained for the membranes analyzed. However, spectra show a similarity on the functional groups found for polypyrrole-cellulose membranes obtained through the same deposition method. There were not important differences on the final composition of membranes regarding the shape used, although there is a variation on the spectra registered regarding the deposition method, which can be attributed to the different concentration of Polypyrrol attached to the cellulose structure due to the procedure used.

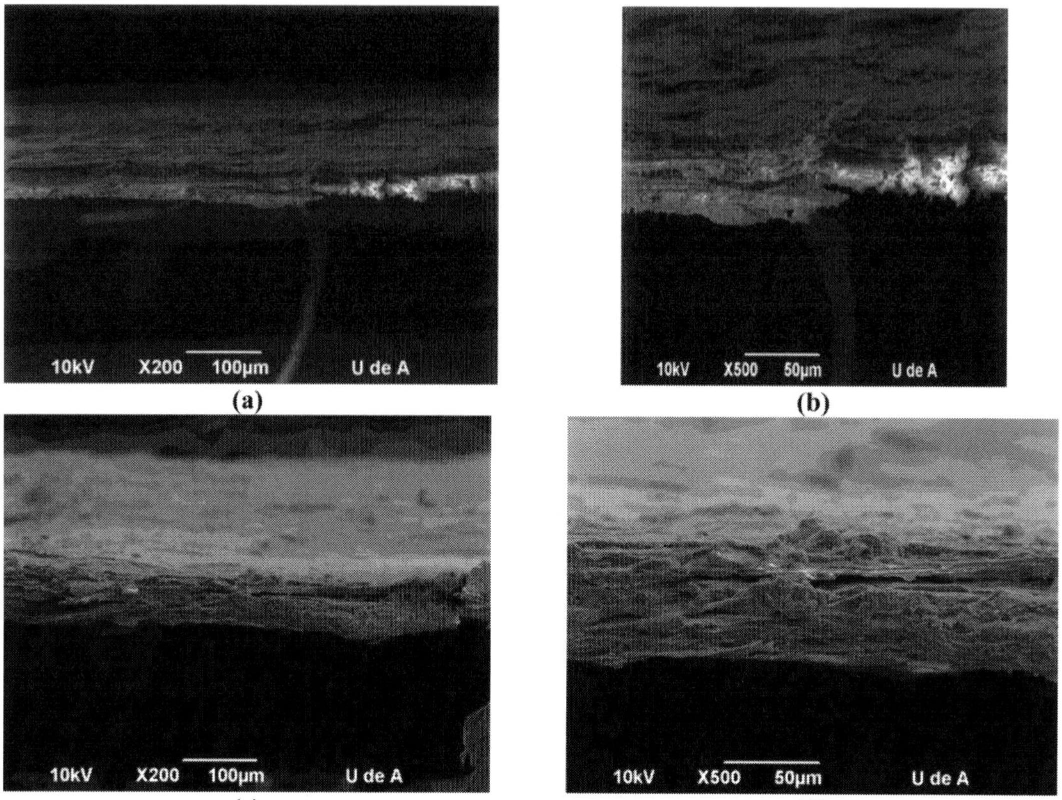

(c) (d)

Figure 2. SEM micrographs of: Images (a) and (b); composite material obtained through direct deposition. Images (c) and (d); composite material obtained through solution deposition.

3.2. Ion removal by the polypyrrole-cellulose membrane

The assay performed to test removal ion application by the composite membranes proved the applicability of the conducting polymer composite on water softening. The reduction on water sample conductivity shows a removal of ions responsible for water hardness. Individual membranes tested achieved conductivity reductions within 23 to 66 µs . cm^{-1} after 6 hours of the assay, which equals a reduction of 11.5 to 33 ppm of CaCO3 for hard water.

On the other hand, variations in the ion removal efficiency among deposition methods and also regarding the membrane shapes used were found. The comparison of efficiency in ion removal for the two deposition procedures used show better results for composite membranes obtained through direct deposition. This variation on the ion removal action may be due to the amount of polypyrrole attached to the cellulose fibers, as was shown in the characterization analysis.

Finally, regarding the initial cellulose membrane shape used to support the polypyrrole; findings show that cylindrical shape does not possess the physical stability needed to be used in water softening. The level of swelling obtainable by the cylindrical shape makes these membranes easy to break and also affects the ion removal application because of the great amount of water in the membrane which reduces the concentration of the ions.

4. Conclusions

The resulting polypyrrole-cellulose membranes combined the electrochemical properties of Polypyrrole and the high surface area of cellulose fibers as expected. The amount of polymer added to the fibers was found to be dependent on the initial concentration of monomer and dopant ion used in the reaction. However, the relation between these two reactants must be further analysed given that at monomer concentrations higher than the dopant ion a harder level of agitation is required in order to obtain a homogenous solution. Concerning to the two different procedure used for the reactants deposition on the fibers, a variation of the characteristics and conductive properties of the membrane was found. Although through solution deposition characterization analysis a more uniform coating of the fibers was obtained, it was through the direct deposition method that membranes showed better ion removal action, given the bigger amount of polymer added to the fibers. Regarding the shapes tested, findings show not only better physical stability for disk shaped membranes, but also better efficiency on ion removal. Finally considering all findings, the Polypyrrole-cellulose membranes obtained through direct deposition from the cellulose are called to be a new material that provides good properties for water softening. Further research should focus on the analysis of the optimal relation between monomer and dopant ion concentration and of the thermal and mechanical properties of the polypyrrole-cellulose membrane.

References

[1] Neves S and Fonseca C 2004 *J. of Brazil Chemestry Society* **15** 395–9
[2] Guimard N, Gomez N and Schmidt C 2007 *Progress in Polymer Sci.* **3** 876–921
[3] Müller D, Rambo C, Recouvreux D, Porto L and Barra G 2011 *Synthetic Metals* **161** 106-11
[4] Merle G, Wessling M and Nijmeijer K 2011 *J. of Membrane Sci.* **377** 1–35
[5] Arsalani N and Geckeler K E 1997 *Reactive and Functional Polymers* **33** 167–72
[6] Johnston J H, Kelly F M, Moraes J, Borrmann T and Flynn D 2006 *Current Appl. Phys.* **6** 587–90
[7] Johnston J H, Moraes J and Borrmann T 2005 *Synthetic Metals* **153** 65–8
[8] Mecerreyes D, Stevens R, Nguyen C, Pomposo J, Bengoetxea M, Grande H 2002 *Synthetic Metals* **126** 173–8.

Journal of Physics: Conference Series **466** (2013) 012013 doi:10.1088/1742-6596/466/1/012013

Comparative study between structural and electrical properties of geopolymers applied to a green concrete

A M Montaño[1,2], C P González[2], J Pérez[1], C Royero[1], D Sandoval[1] and J Gutiérrez[1]
[1] Escuela de Química, Universidad Industrial de Santander, Bucaramanga, Colombia.
[2] Grupo de Investigación en Materiales - GIM, Departamento de Ciencias Básicas, Universidad Pontifica Bolivariana Seccional Bucaramanga, Floridablanca, Colombia.

E-mail: claudia.gonzalez@upb.edu.co

Abstract. This work shows a comparative analysis of geopolymers obtained by alkaline activation of two aluminosilicates: bentonite and metakaolin. With the goal of to replace some cement percentage, both aluminosilicates were added in several proportions (10, 20 and 30%) to concrete mixes. Portland Type I cement was used to prepare the reference concrete (without geopolymer). X-ray diffraction of geopolymers allowed to find new crystallographic phases that was not present in precursor's minerals. To evaluate mechanical properties of concrete prepared with geopolymers, test tubes with 7, 14, 28 and 90 days as setting time were used. Chemical resistance and Electrical impedance of concrete mixes were also measured. Results shows that cementitious material obtained from metakaolin exhibit the best compressive strength. On the other hand, those materials derived from bentonite, have a high electrical resistance so that, they protected reinforced concrete better that Portland does.

1. Introduction

Portland type I cement (PC), is the main component of concrete because of the high resistance that it offers in hardened state. Nevertheless, cement manufacturing, has remarkable consequences on environmental contamination due to the emission enormous during the process of obtaining Clinker. For that reason to search alternatives materials for cement it is necessary. Inorganic polymers, also called Geopolymers [1-5], are very interesting materials for both investigation and building industry for its mechanical resistance similar to that for Portland, representing one possibility for replacement. This study presents a comparative analysis of two geopolymers obtained by alcaline activation of betonite (BEN) and *meta*caoline (MEC). This geopolymers were added as cement replacement in different concentrations to concrete mixes.

2. Experimental

Geopolymers synthesis starts by aluminosillicate (BEN and MEC) addition to alkaline solution of Na_2SiO_3. Mix reaction was bring to 60-120 °C by 31 h and dried by 6 h, to obtain cementitious materials. Subsequently, geopolymers was ground to a particle size of 4,75 mm (according to ASTM C94 /C 94M – 09) [6] to elaborate concrete test tubes.(according to ICONTEC NTC 1377) [7] with Portland substitution of 10, 20 y 30%. It was prepared cylinders 75 mm in diameter and 150 mm high, for mechanical testing of each mixture. Test pieces of 2 cm diameter and 4 cm in length, was made for an electrochemical analysis [5]. The setting times employed was 7, 14, 28 and 90 days.

Journal of Physics: Conference Series **466** (2013) 012013 doi:10.1088/1742-6596/466/1/012013

3. Results and discussion

BEN geopolymer diffractogram (figure-1) shows new crystalline phases which are not presents in bentonite: natrolite, microcline, gibbsite and corundum. These phases are coming from alluminosilicate structure transformation because of the alkaline solution. There are some phases that are conserved: illite, quartz and montmorillonite which showed a 0.5° shift at 2θ. Meanwhile, MEC geopolymer (figure 2) it has not a crystal structure this is an amorphous material with some phases like muscovite, anatase, microcline and quartz from kaolin starting material.

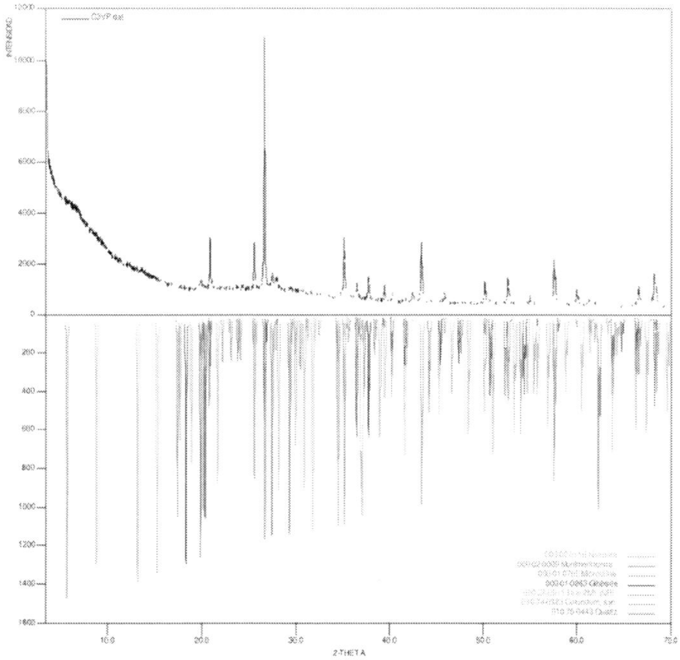

Figure 1. X- ray diffraction patterns of the BEN geopolymer.

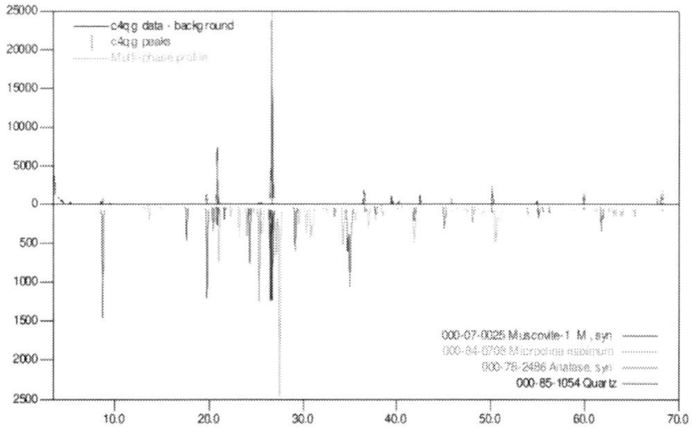

Figure 2. X- ray diffraction patterns of the MEC geopolymer.

Test to compressive strength of concrete were conducted. Test specimens which was replaced at 10, 20 and 30% Portland cement by geopolymer and reference (without replacement) were subjected to a setting of 7, 14, 28, 90 and 120 days. In figure 3 and 4 shows compressive strength for concretes with BEN geopolymers (GP BEN) and MEC geopolymers (GP MEC) respectively.

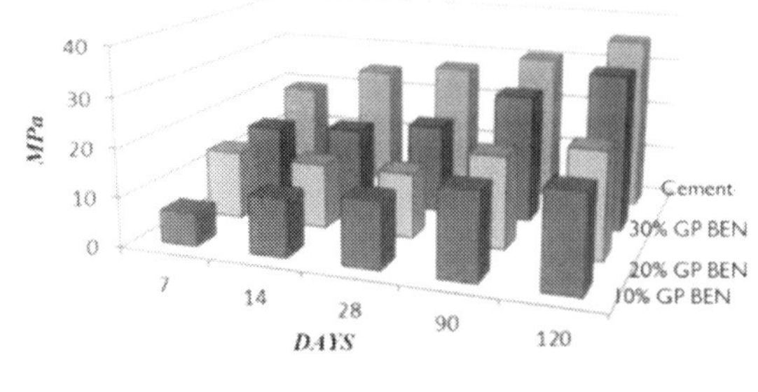

Figure 3. Compressive strength of the specimens with substitution of BEN geopolymer and reference probe.

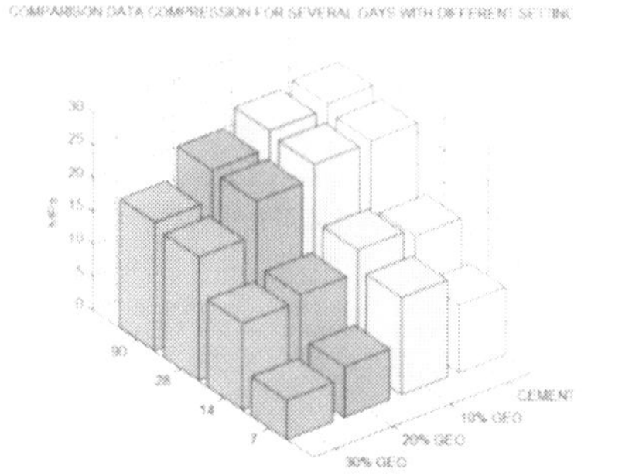

Figure 4. Compressive strength of the specimens with substitution of MEC geopolymer and reference probe.

For both concretes it is found that the best compressive strength corresponds to reference. However, in the samples with 30% GP BEN resistance value was higher than in the other replacement percentage. It can also be observed that as the curing time increases, also increases the resistance value, showing an almost linear behavior because, in concrete, the hydration reactions take place at high speed. High strength values for the samples tested with 120 days of curing with and without replacement of Portland. Similar behavior was observed in the specimens with GP MEC.

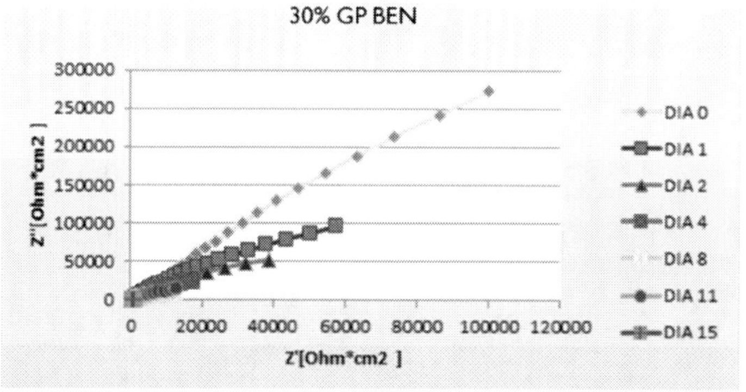

Figure 5. Nyquist graphs obtained for the specimen with 30% GP BEN.

Nyquist impedance spectrum at different times and different immersion substitutions geopolymer were performed. It was observed for all diagrams same tendency to increase the time of analysis, the value of resistance decreases, thereby increasing the corrosion process that occurs when exposing the steel to the 2% NaCl solution. Figures 5 and 6 show Nyquist diagrams of the impedance spectrum at different immersion times for the samples with 30% GP BEN and 10% GP MEC respectively. In the sample with 30% GP BEN, corrosion is less due to there is less charge transfer to the metal-mixture interface, thus presenting different electrochemical behavior for each mixture and setting time. Between 8 and 11 immersion days it was an increase in the impedance value then decreased again to the 15th measurement, following the same corrosive behavior that was submitted over time. This change in behavior of the material in time, is attributed to the oxide layer formed on the surface of the steel is shield type (which is an oxide that reduces the exchange of species from the solution to the material).

A comparable behavior in the test tubes was presented with 10% of the GP MEC for 11 days but with lower impedance values with respect to the GP BEN.

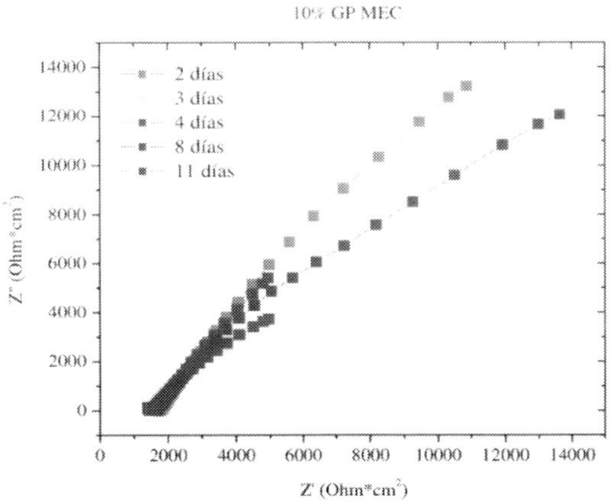

Figure 6. Nyquist graphs obtained for the specimen with 10% de GP MEC.

4. Conclusions

Cementitious materials formed from metakaolin showed the best mechanical resistance, while those made from bentonite, exhibited a high electrical resistance so that there was a greater protection against corrosion in reinforced concrete.

Electrical tests confirm that the optimum percentage of geopolymer MEC replacement for cement in a concrete mix should be 10%, since in these measures were reached high impedance value, indicating the increased electrical resistance.

References

[1] R Mejía de Gutiérrez, S Bernal and E Rodríguez 2006 *Congreso Iberoamericano de Metalurgia y Materiales* (Cuba: Habana)
[2] Narayan H, Alemu H and Iwuoha E 2006 *Physics Status Solidi.* **203-15** 3665
[3] K Komnitsas and D Zaharaki 2007 *Minerals Engineering* **20** 1261
[4] Dorith Tavor, Adi Wolfson, Anat Shamaev and Asia Shvarzman 2007 *Ind. Eng. Chem. Res* **46** 6801
[5] H Narayan, Angela M Montaño, Monica L Hernández, July A Hernández, Claudia P González and César A Ortiz 2012 *J. Mater. Environ. Sci.* **3-1** 137
[6] ASTM C94 /C 94M 2009 *Standerd Specification for Ready-Mixed Concrete* (Estates Unites: American Society of Testing and Materials - ASTM)
[7] NTC 1377 1994 *Ingeniería civil y arquitectura, Elaboración y curado de especímenes de concreto para ensayos de laboratorio* (Colombia, ICONTEC)

Journal of Physics: Conference Series **466** (2013) 012014 doi:10.1088/1742-6596/466/1/012014

Effect of coatings obtanied by sputtering of chromium catode on the corrosion resistance of AISI H13 steel

A Sandoval[1], D Peña[1] and U Piratoba[2]
[1] Grupo de Investigaciones en Corrosión - GIC, Universidad Industrial de Santander, Bucaramanga, Colombia.
[2] Grupo de Superficies Electroquímica y Corrosión, Universidad Pedagógica y Tecnológica de Colombia.

E-mail: anderson84f@hotmail.com

Abstract. Corrosion resistance of coatings obtained by sputtering a chromium target were evaluated. The films were deposited on substrates of disk-shaped AISI H13 steel. By means of potentiodynamic polarization curves were able to determine the current density vs. potential for the coated and uncoated substrate and the difference in the corrosion potential Ecorr. All samples with coating showed an increase in Ecorr respect to substrate. The electrochemical tests were conducted in an electrolytic solution of 3% NaCl.

1. Introduction

To carry out the corrosion behavior, it has been implemented in some direct current electrochemical techniques such as polarization resistance or linear polarization and potentiodynamic polarization curves [1-3], also techniques of alternating current are used as electrochemical impedance spectroscopy (EIS) [4-6]. The potentiodynamic polarization curves or Tafel curves are used to characterize different types of metals, this technique generates a semilog plot where the polarization curves are located. Tafel technique is a destructive technique from interface point of view since it gets it messy, which causes the movement of ions which changes the state of equilibrium of the interface [1, 7, 8].

2. Experimental

The material used was AISI H13 steel, whose chemical composition is: 0.39% C, 1.0% Si, 0.4% Mn, 5.3% Cr, 0.9% V, 1.3% Mo [8]. This steel was selected because it is suitable for nitriding and coated by PVD. This steel provides good balance of toughness, high resistance to cracking caused by thermal shock and resistance to tempering, combined with moderate wear resistance.

The evaluation of the corrosion resistance of the samples was carried out with a Gamry Potentiostat Model PC4. The system consists on a flat electrochemical cell, incorporating Ag / AgCl as reference electrode and a platinum electrode as counter electrode. The electrolytic solution used for the tests was 3% NaCl. The polarization Tafel curves were carried out by scanning the potential between -300 mV and 300 mV around the equilibrium potential with a rate of 1 mV/s.

3. Results and discussion

The films studied presented different colors, M1 yellow, M2 Coffe, M3 and M4 dark gray.

The figure 1 shows the potenciodynamic polarization curves obtained for different samples, you can observe that for all films the corrosion potentials were more positive than the substrate, which enables to state that its thermodynamic stability was enhanced. The corrosion currents results shows lower values to the films than the uncoated material, which corroborates the values obtained in the corrosion rates, which are lower for steel coating than uncoated steel, with these results it can be concluded that this treatment improves the performance of the test pieces front of corrosive processes.

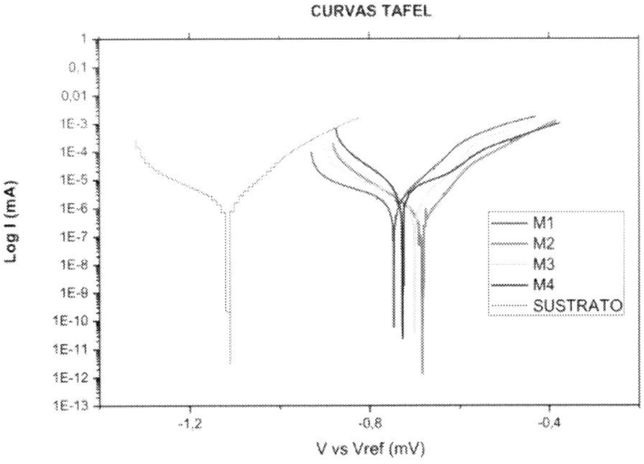

Figure 1. Potenciodynamic polarization curves.

Table 1. Parameters obtained from the Tafel curves.

Parameter	Ecorr (mV)	Icorr (μA)	Corrosion Rate (mmpy)
Substrate	-1110	1.53	7.98E-01
M1	-747	0.692	3.53E-01
M2	-683	0.21	1.07E-01
M3	-701	1.13	5.75E0.1
M4	-727	0.353	1.80E-01

4. Conclusions

The results show that all coatings obtained allow to increase the corrosion resistance of steel AISI H13, as able to increase the corrosion potential of about 300 mV and decrease the corrosion current in all cases. Although the Tafel curves obtained showed no passivation areas, it was observed that the treatment improves the properties of the steel.

References

[1] Enrique Vera López 2000 *Curso Básico De Corrosión Protección Catódica y Recubrimientos* (Colombia: Universidad Pedagógica y Tecnológica de Colombia)
[2] Carlos Martínez, Yenny Gonzalez, E Vera, Cesar Peña 2009 *Revista Colombiana de Tecnología Avanzada* **1-13** 33
[3] J J Olaya, S E Rodil, S Muhl, E Sanchez 2005 *Thin Solid Films* **474** 119
[4] D D Macdonald 1990 *Corrosion* **46-3** 229
[5] Darja Kek Merl, Peter Panjan, Mihan Cekeda, Marijan Macek 2004 *Eltrochemica Acta* **49** 1527
[6] Valentina Millano, Miguel Sánchez, Oladis T de Rincón, Douglas Linares y Nathalie Romero 2006 *Revista Técnica de la Facultad de Ingeniería Universidad del Zulia* **29-3** 291
[7] V A Alves et al. 2005 *Corrosion Science* **47** 2871
[8] H J Flitt, D P Schweinsberg 2005 *Corrosion Science* **47** 3034

Journal of Physics: Conference Series 466 (2013) 012015 doi:10.1088/1742-6596/466/1/012015

Temporary spectral analysis of a laser plasma of mineral coal

P Rebolledo[1], P Pacheco[1], R Sarmiento[1], R Cabanzo[2] and E Mejía-Ospino[2]
[1] Grupo de Espectroscopía Óptica de Emisión y Laser - GEOEL, Universidad del Atlántico, Barranquilla, Colombia.
[2] Grupo de Espectroscopía Atómica y Molecular - LEAM, Universidad Industrial de Santander, Bucaramanga, Colombia.

E-mail: pablorebolledo@mail.uniatlantico.edu.co

Abstract. In this work we present results of the temporal spectral study of a plasma laser of mineral coal using the Laser-induced Breakdown Spectroscopy (LIBS) technique. The plasma was generated by focusing a laser beam of Nd:YAG laser emitting at 532 nm with energy per pulse of 35 mJ on coal target pellets. The plasma radiation was conducted by an optical fiber to the entrance slit of a spectrograph of 0.5 m, equipped with a 1200 and 2400 grooves/mm diffraction grating and an ICCD camera for registration with different delay times of the spectra in the spectral range from 250 nm to 900 nm. The temporal spectral analysis allowed the identification of the elements Al, Fe, Ca, Mg, K, and Si, and CN and C2 molecules present in natural coals. The characteristics of the spectral lines and bands were studied at different delay times obtaining the calculation of the evolution of electron temperature, electron density, and vibrational temperature of plasmas in the time. The delay times used were between 0.5 μs and 5 μs, calculating the electron temperature ranged between 5 000 K and 1 000 K.

1. Introduction

The mineral coal together the petroleum and the natural gas are at present the biggest sources to energy production useful worldwide. Most of the energy consumed worldwide comes from these sources, but associated to them it is generated a big environmental contamination because of the emission of CO_2 and other gases. The mineral coal is compound by carbon, hydrogen, oxygen and a traces S, Al, Fe, Hg, Mg, Na, Ca, V, Mn, K and other characteristic elements of the inorganic matter [1, 2]. Some of these elements can vaporize during the coal burning, while others persist in the ashes. In this sense, it is necessary the continuous monitoring during all the process that include the exploitation, transportation and use of the coal in the plants. The LIBS technique is a powerful method to the detection and quantification of chemistry species in solid samples, liquid and gaseous [3, 4]. In this work it has been implemented and applied the LIBS technique to the identification of polluting substances in three coal samples coming from the Colombia Caribbean Region. With the spectral results that have been obtained and the use of atomic and molecular parameters reported [5, 6], it was done an estimation of the electronic temperature and density, just like the vibrational temperature of lasers plasma coal. It is also reported a temporal analysis of the integrated intensity of emission lines just like the evolution on time of the temperature and the electronic density.

2. Experimental description

To the temporary and spectral analysis of the mineral coal plasmas there were made pellets in labs following the norm NTC 3266 [7]. The pellets were made by compacting crushed coal to a pressure of

12 000 PSI during 10 minutes without any kind of agglutinative getting pellets of dimensions of 0.016 m of diameter and 0.004 m of thickness. The plasmas laser coals were generated in air atmosphere, and in argon atmosphere, focusing the radiation of a laser of Nd: YAG operating in 1064 nm and 532 nm with an average energy per pulse of 35 mJ and to a frequency of 5 Hz. The radiation emitted from the plasma was focused with a quartz lens of 0.05 m of focal distance over the extreme of an optical fiber which conducted it until the entrance slit of an optical spectrograph of 0.5 m of focal distance, equipped with three diffraction grating of 600, 1200 and 2400 grooves/mm and with an ICCD camera. The spectra were registered in the spectral region of 250 nm to 900 nm, with different delay times. The data were processed by using commercial software and to the assignation of ionic and atomic lines and molecular bands it was consulted the data basis, and molecular parameter tables [5, 6]. Figure 1 shows a basic diagram of the experimental assembly of the LIBS technique resolved in time, to the generation of plasma laser and to the temporary acquisition of the spectra.

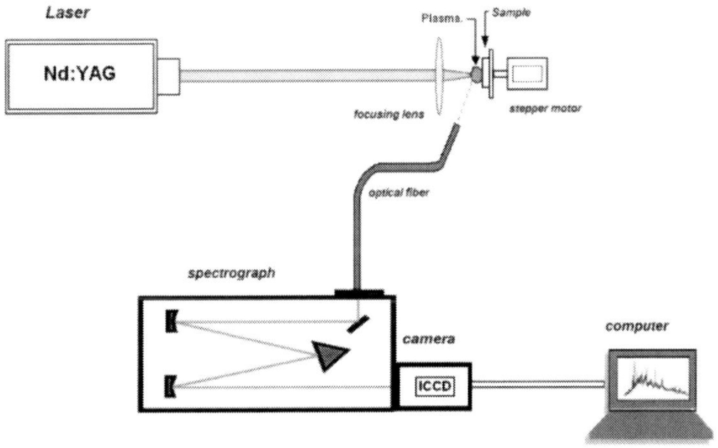

Figure 1. Diagram of the LIBS assembly resolved in time.

3. Results and discussion

The spectral analysis of plasma laser generated in the three coal samples of Calenturitas, Cerrejon and La Loma mines let to establish the presence of the elements Fe, Na, K, Al, Mg, Ca, Si, H, O, and N, how it is shown on figure 2 with centered windows in 290 nm figure 2 (a) and 390 nm figure 2 (b), took with a delay time of 0.5 μs. On the three spectra of the Figure 2a are observed ionic lines in 279.564 nm and 280.270 nm of Mg II, and atomic lines in 285.215 nm of Mg I, 288.214 nm of Si I and 302.103 nm of Fe I, while in the spectra of the figure 2b it is observed the band of CN of the violet system with band heads in 388.246 nm (0-0); 387.059 nm (1-1); 386.069 nm (2-2); 385.419 nm (3-3); 384.995 nm (4-4), and ionic lines of calcium in 393.364 nm and 396.814 nm of Ca II, the aluminum atomic in 394.410 nm and 396.163 nm of Al I, and the iron line in 404.583 nm of Fe I.

It was also studied the temporal evolution of the integrated intensities of atomic lines, ionic lines and molecular bands, to delay times included between 0.5 μs and 5.5 μs. It was done a graphic of integrated intensity in function of the delay time observing that the intensities low with the time. The figure 3 (a) shows the variation of the intensity of the lines 279.552 nm of Mg II and 285.205 nm of Mg I; it is observed that at 0.5 μs both lines show big intensity because of that this delay time is considered optimal to take the spectra in this region. The temporary studio of the band intensity of CN with transition $B\ ^2\Sigma^+ \rightarrow X\ ^2\Sigma^+$ was done by measuring in different times the integrated intensity of the band head in 421.753 nm (0-1).

Journal of Physics: Conference Series **466** (2013) 012015 doi:10.1088/1742-6596/466/1/012015

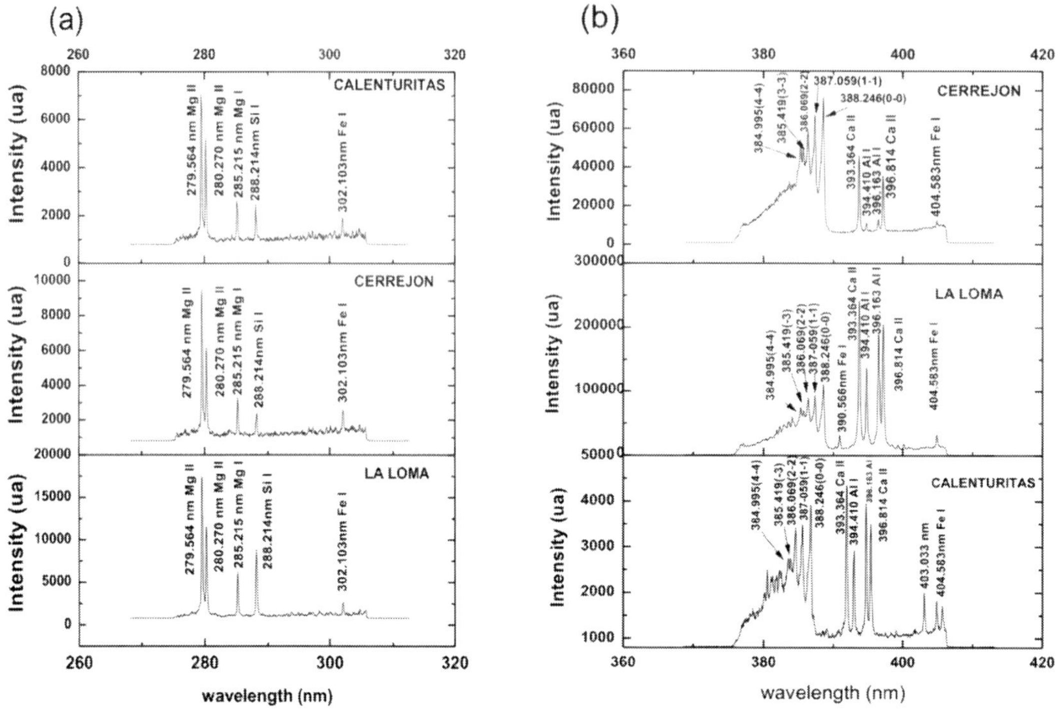

Figure 2. LIBS spectra of three coal samples.

The plasma studio generated in argon atmosphere showed that this atmosphere increases the intensity of C_2 bands while it lows the intensity of the CN bands, this last incident benefits the observation of spectral lines apparently of low intensity how we can see on figure 3 (b), where the atomic lines of iron 398.856 nm, 399.779 nm and 400.451 nm are only visibles in the spectrum took in argon atmosphere (lower) and they don´t appear in the spectrum took in the air atmosphere (top).

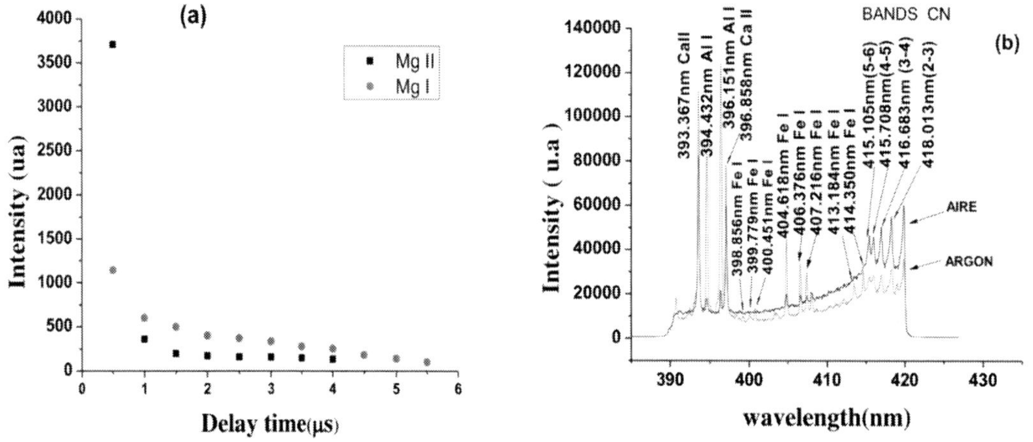

Figure 3. (a) Variation in the integrated intensity time of atomic and ionic lines (b) spectrum of coal plasma laser generated in argon atmosphere.

The characterization of the coal plasmas was done by applying diagnostic methods of plasma optical spectroscopy to the determination of the electronic temperature, the electronic density and the vibrational temperature under the consideration of Local Thermodynamic Equilibrium (LTE). Figure 4 shows the behaviour of these parameters on the time, it is observed that as increases the delay time decreases each of these parameters due to the process of cooling the plasma.

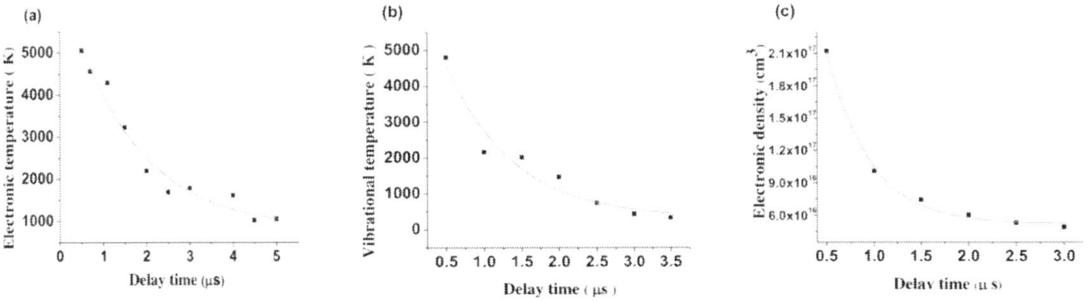

Figure 4. (a) Electronic temperature variation on the time, (b) Vibrational temperature variation on the time, (c) Electronic density variation on the time.

The calculus of the electronic temperature of plasmas was done by means of the Boltzmann's plot by using the lines 427.123 nm, 430.820 nm, 432.576 nm and 438.342 nm of Fe I, estimating values between 5 000 K and 1 000 K, to delay times between 0.5 μs y 5 μs. To the estimation of the electronic density were used the atomic parameters of the isolated line 288.204 nm of Si I giving values between 2.12×10^{17} cm^{-3} and 5.00×10^{16} cm^{-3}.

4. Conclusions

It was implemented the LIBS technique resolved in time and it let to detect the elements Al, Fe, Na, Ca, Mg, K and Si in mineral coal samples of mines of the Colombia Caribbean Region. With the spectra temporal measures it was done a diagnostic of the plasmas generated in different atmospheres. The results also let to estimate the electronic temperature and density, and the vibrational temperature, and the delay time appropriated to the optimization by the acquisition of the spectra. It was analyzed the variation of the bands and lines intensities when the spectrum is generated in argon atmosphere.

References

[1] Mateo M P, Nicolas G, Yanez A 2007 *Appl. Surf. Sci.* **254** 868
[2] Jie L, Jidong L, Zhaoxiang L, Shunsheng G, Chengli X, Liang Ch, Lifei Y and Pengyan L 2009 *Optics and Laser Technology* **41** 907
[3] Miziolek A, Palleschi V and Schechter I 2006 *Laser–Induced Breakdown Spectroscopy: Fundamentals and Applications* (New York: Cambrige University Press)
[4] Radziemski L and Cremers D 1989 *Laser–Induced Plasma and Applications*, (New York: Marcel Dekker)
[5] NIST: http://www.nist.gov/pml/data/asd.cfm
[6] Rosen B 1970 *Selected Constant spectroscopy data relative to diatomic molecules* (New York: Ed. Pergamon Press)
[7] NTC 3266 2010 *Muestreo y preparación de muestras de coque para análisis de laboratorio* (Colombia ICONTEC)

Correlation between the OES plasma composition and ZnMnO films properties during pulsed laser

H Riascos[1], D Ramírez[1] and J Uzurriga[1]
[1] Plasma, Láser y Aplicaciones, Universidad Tecnológica de Pereira, A. A. 097, Pereira, Colombia.

E-mail: hriascos@utp.edu.co

Abstract. Manganese doped zinc oxide was ablated on (100) silicon using the fundamental harmonics of Nd:YAG laser at various oxygen ambient gas pressures and temperature substrate was kept at room temperature during the deposition. In all experiments we used ZnMnO target. Optical emission spectroscopy (OES) is used to survey the plasma composition during deposition. The intensities of the ZnI and ZnII emission lines are correlated to the oxygen gas pressure and composition and morphology of the films deposited. Strong emission lines of ZnI, Zn II and neutral oxygen are observed. Oxygen and Mn emission lines in its atomic or ionic states have not been detected in this investigation, which is thought ZnMnO as being exclusively produced on the substrate surface and not generated into the plasma, which is in agreement with previous studies of ZnO plasma plume. Atomic Force Microscopy (AFM) is used to study the influence of oxygen gas pressure on the ZnMnO film morphology. An X-ray diffraction was tried to verify a crystal structure of the sample. We have demonstrated that gas pressure is an important parameter for the quality and performances of ZnMnO structures obtained by pulsed laser deposition.

1. Introduction

Pulsed Laser Deposition (PLD) of material processing is a technique which has given good results for various applications of thin films of wide range materials [1]. In PLD, a pulsed laser is focused onto a target of the material to be deposited. For sufficiently high laser energy density, each laser pulse vaporizes or ablates a small amount of the material creating a plasma plume. The ablated material is ejected from the target in a highly forward-directed plume. The ablation plume consists primarily of excited atomic, diatomic, ionic and others species. The plasma properties of the plume largely determine the quality of the thin films deposited on the substrate. Most of the work about PLD is now devoted to improve the quality of the deposited films. To this end, it is of great importance to understand the ablation process and the transport phenomena of materials from the target to the collecting substrate. The stream of atoms, ions, molecules, and clusters ejected from the target surface emit radiation that can be analyzed using optical emission spectroscopy to gather information about the composition and dynamics of the plasma plume.

ZnO is a semiconductor possessing a large number of interesting properties that have been the focus of intensive research for many years. Due to recent and exciting developments in several key areas, pure or doped ZnO films appear to be a highly promising new and advanced functional material. Initially interest in oxide based in semiconductors was stimulated by the theoretical prediction of Dietl et al [2] that ZnO would exhibit ferromagnetism at room temperature after doping with Mn. Until now several research groups attempted to obtain experimental evidence for such ferromagnetism in thin films form of ZnO doped with transition metals. Thin Films of ZnO and ZnO-related compounds have been grown by a wide a variety of chemical as well as physical methods: sol-gel synthesis [3], chemical vapour deposition [4], magnetron sputtering [5], molecular beam epitaxy (MBE) [6] and Pulsed Laser Deposition [7].

In this work we report the analysis of plasma generated by pulsed laser ablation using fundamental wavelength (1064 nm) of Nd:YAG (neodymium-doped yttrium aluminum garnet) laser. The effect of O_2 gas pressure on plasma plume and some thin films were studied comparatively.

2. Experimental methods

The plasma plume was produced by ablating a ZnMnO target using a Nd:YAG laser (wavelength: 1064 nm, pulse width: 9 ns, repetition rate: 10 Hz) at various oxygen ambient pressures from 13 to 95 mTorr. The ZnMnO target was mounted on a holder rotated at a constant rate of 2.2 rpm so that the laser beam continuously irradiated a fresh region on the surface. The laser beam was focused onto the target at an incident angle of 45° with respect to the target surface at power of 1.2 W. The optical emission spectra from plume were measured using a spectrometer (Jobin Yvon Triax550) with a wavelength resolution of 2 nm and 1200 grooves/mm grating attached to a CCD camera. The ablation was done in O_2, atmosphere at pressures ranging from 13–95 mTorr.

The surface morphology and crystallite size of ZnMnO in the films were recorded by atomic force microscopy (AFM). Also, other properties of thin films were observed by X-ray diffraction (XRD).

3. Analysis of the plasma

To investigate the plume dynamics of the ejected material accompanying laser ablation of the ZnMnO target, OES measurements were carried out, obtaining information on the nature of the ejected species. The optical emission spectra of plasma plume were recorded in the wavelength range from 400 to 800 nm at various gas pressures and substrate temperature.

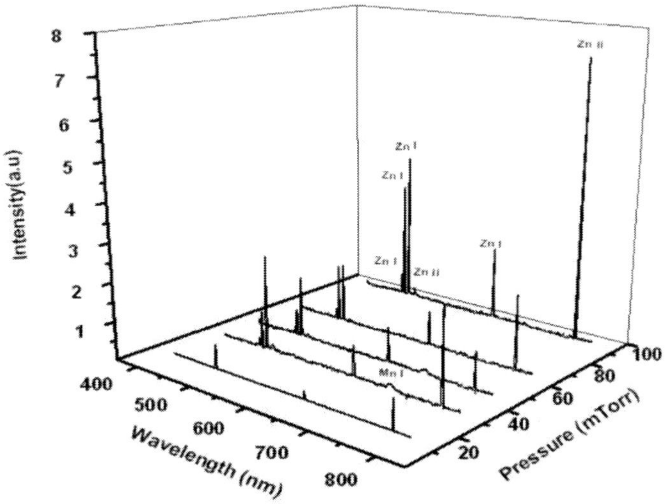

Figure 1. Typical optical emission spectra of the plume accompanying the 1064 nm pulsed laser ablation of the ZnMnO target in oxygen atmosphere at a pulse power of 1.2W.

Strong blue light emission from the plasma due to the emissions from the neutral Zn atoms could be observed. Typical emission spectra of the plasma plume formed during laser irradiation in an O2 atmosphere and at a laser power 1.2 W are shown in figure 1. These spectra showed that the plasma emission in mainly due to atomic and singly ionized Zn. These results are similar to those previously reported for pulsed-laser ablation of ZnO [8–10] and, Saji et al [11] who only observed zinc atoms and ions during the ablation of a sintered ZnO target by Q-switched third harmonic Nd:YAG laser (355 nm) at different oxygen partial pressures. Strong emissions were observed around 481.05 and 636.23 nm from the neutral Zn atoms (Zn I) and 773.25 from the singly ionized Zn atoms (ZnII) in the all gas pressure, others neutral spectral lines such as 468.01, 472.1, and the ionic line 492.4 appear in all pressures except at 13 mTorr. Practically at 13 mTorr there are a few presences of Zn emission lines.

The peaks of the neutral and ionized Mn atoms and O species could not be observed, possibly due to a lower density of these compared with that of the neutral Zn atoms. On the other hand, during ablation, the ionization potential of atomic Zn is much lower than those of the other atomic species in the plume (9.39 eV for atomic Zn, 13.62 eV for atomic O), [9]. Therefore, Zn species are the dominant in the dense plasma. Significant variations in the intensities of emission lines of various atomic and ionic species of the plasma were observed with changes ambient gas pressures, it is observed that both electronically excited Zn atoms and the singly ionized Zn atoms peaks increase linearly with gas pressure as shown in figure 2. The intensities of the emission lines corresponding to the Zn I and Zn II reach maximum values at 95 mTorr. The intensity increase of the spectral lines is due to the effect of cooling by the background gas resulting in confinement of the plasma plume. The strongest observed emission lines are listed in table 1 with corresponding spectroscopic data. From figure 2, it is seen that the emission intensity of strongest lines increase with Oxygen gas pressure.

Table 1. The strongest observed emission lines of the plume accompanying the 1064 nm pulsed laser ablation of the ZnMnO target in oxygen atmosphere at a pulse power of 1.2W.

	Lines ZnO	
ZnII	492.4 nm	$4d^2D_{5/2}$-$4f^2F_{7/2}$
ZnI	468.01 nm	$4s4p^3P_0$-$4s5s^2S_1$
ZnI	472.21 nm	$4s4p^3P_1$-$4s5s^3S_1$
ZnI	481.05 nm	$4s4p^3P_2$-$5s^3S_1$
ZnI	636.23 nm	$4s4p^1P_1$-$4s4d^1D_2$

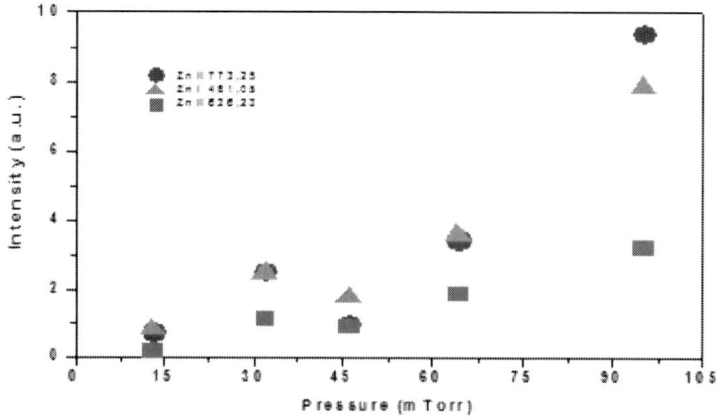

Figure 2. (color online) Variations of Intensity emission lines of the plume corresponding to ZnI (481.05 nm, 636.23 nm) and ZnII (773.25 nm) with increase of oxygen pressure.

4. Deposition of ZnO films

ZnMnO thin films were deposited on silicon (111) substrate at room temperature, various gas pressure from 13 to 95 mTorr by PLD using a Nd:YAG laser with 1.2W per pulse. The focused laser radiation illuminated a rotating target at an incident angle of 45°. The target was 99.9% pure ZnO doped with 0.05wt% Mn. The distance target-substrate was 5.5 cm.

4.1. XRD analysis

The ZnMnO thin films grown at room temperature on various oxygen gas pressures have been characterized with regards to the structure by XRD. Amorphous films were obtained, except the film grown at 95 mTorr as shown in figure 3. This suggests that the ablated species reaching the substrate may not have the minimum energy required for crystallization and thereby getting amorphous films. The X-ray diffraction peak corresponding to ZnO (002) plane is observed. Observation of the (002) plane indicates that the film is c-axis oriented. This indicate that the surface energy of (002) plane is the lowest in ZnO crystal [12].It is known that the number of particulates arriving at the substrate is decreased as the pressure of an ambient gas increases. Deposited particulates in a ZnO film degrade film quality. Therefore, the crystal quality is improved as the ambient gas pressure increase [13].

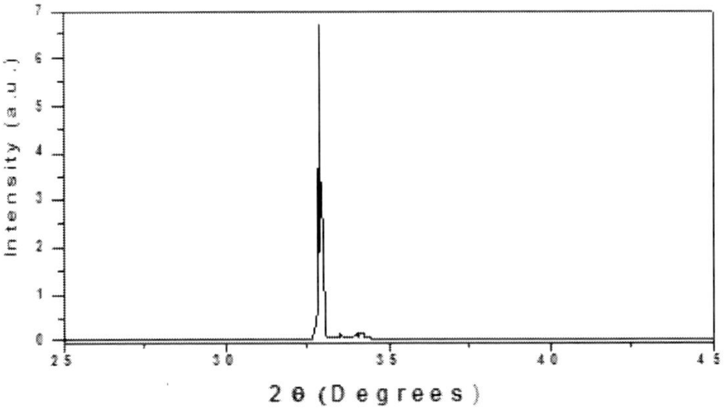

Figure 3. XRD 2θ scan of film deposited on Silicon (100) substrate at room temperature for 95 mTorr oxygen pressure by PLD.

4.2. Surface morphology by AFM

Films deposited under different oxygen pressures at room temperature were also evaluated by AFM. The scanning area was 1 μm×1 μm.

Figures 4a) to 4e) show the surface morphology of these samples as seen by AFM. The images show the effect of gas pressure on the surface roughness and grain size. It is seen that the grain size increase with oxygen gas pressure, with a grain-size distribution between 15.9 and 44.6 nm for the samples grown at room temperature. In contrast, the roughness behaves very different; there is no a clear trend of increasing roughness with increasing gas pressure.

Table 2. Structural parameters for the ZnMnO samples obtained from AFM at RT.

Oxigen Pressure (mTorr)	rms surface roughness (nm)	Grain size (nm)	Thickness (nm)
13	0.55	15.89	65.9
32	2.87	19.90	360.0
46	1.20	26.57	97.8
64	2.74	40.16	267.0
95	3.92	44.59	180.0

Table 2 summarizes the results of AFM measurements on ZnMnO thin films grown at different pressures. The results can be understood by considering the ZnMnO plume size. As the ambient pressure increases, the number of particles arriving on the substrate decreases, resulting in deterioration of the deposited film.

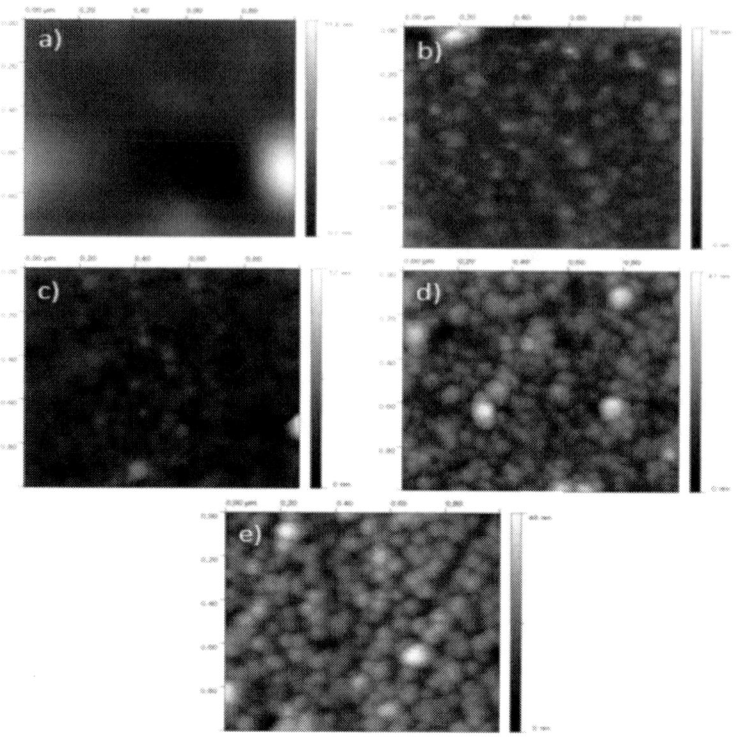

Figure 4. (color online) Images of ZnMnO films prepared in oxygen atmosphere at various pressures and room temperature by Nd:YAG (1064 nm, 9 ns, 1.2W, 10 Hz) on silicon substrate.

5. Conclusions

We observed that only at 95 mTorr ZnMnO crystalline films was grown. Optimization of the deposition conditions makes it possible to produce high-quality thin films. Structure and surface morphology of all obtained samples demonstrated the strong influence of the gas pressure. OES investigations are in agreement with the studies of films, showing plasma richer in Zn species Our studies evidenced the prevalent presence of ZnI in the plasma generated under action Nd:YAG laser operated at 1064 nm.

References

[1] Robert Eason 2007 *Pulsed laser deposition of thin films: applications-led growth of functional materials* (A Wiley-Interscience publication)
[2] T. Dietl, H. Ohno, F. Matsukura, J. Cibert, D. Ferrand 2000 *Science* **287** 1019
[3] Y M Kim, M Yoon, I-W Park, J H Lyou 2004 *Solid State Commun.* **129** 175
[4] P F Zhang, H Y Wei, G W Cong, W G Hu, H B Fan, J J Wu, Q S Zhu, X L Liu 2008 *Thin Solid Films* **516** 925
[5] H Li, X J Pan, M Qiao, Y Z Zhang, T Wang and E Q Xie 2008 *Vacuum* **82** 459
[6] A Bakin, A El-Shaer, A C Mofor, M Kreye, A Waag, F Bertram, J Christen, M Heukena, J Stoimenos, J Cryst 2006 *Growth* **287** 7
[7] R O'Haire, E McGlynn, M O Henry, J P Mosnier 2007 *Superlattice Microst.* **42** 468–72
[8] R K Thareja, H Saxena and V Narayanan 2005 *Appl. Phys.* **98** 034908
[9] F Claeyssens, A Cheesman, S J Henley, M N R J Ashfold 2002 *Appl. Phys.* **92** 6886
[10] S Namba, R Nozu, K Takiyama and T Oda 2006 *Appl. Phys.* **99** 073302
[11] K J Saji, R Manoj, R S Ajimsha and M K Jayaraj 2006 *Proc. of SPIE* **6286** 62860D
[12] K L Chopra, S Major and D K Pandaya 1983 *Thin Solid Films* **102** 1
[13] Y R Ryu, S Zhu, S W Han and H W White 1998 *Vac. Sci. Technol.* A **16** 3058

Journal of Physics: Conference Series **466** (2013) 012017 doi:10.1088/1742-6596/466/1/012017

Preparation of dental amalgam and spectral diagnosis of mercury in plasmas-laser in the region of 250 nm - 850 nm

A De la Ossa[1], **P Pacheco**[1] **and R Sarmiento**[1]
[1] Grupo de Espectroscopía Óptica de Emisión y Laser – GEOEL, Universidad del Atlántico, Barranquilla, Colombia.

E-mail: paolapacheco@mail.uniatlantico.edu.co

Abstract. In this paper we presents results of the spectral study of plasmas-laser of dental amalgam by technique Laser-induced Breakdown Spectroscopy (LIBS). Plasmas were generated focusing the beam of a Nd: YAG laser on the matrix of the mixture Ag-Sn-Cu and on amalgams with different proportions of mercury (3:2, 5:2, 6:2). Based on the spectral results and reported atomic parameters, became estimation of electron temperature plasmas- laser and their behavior with the concentration of Hg. The estimated values of the electron temperature for the respective proportions were 20 846 K, 19 139 K and 16 872 K, using the distribution of population of Boltzmann energy levels associated with spectral lines, considering conditions Local Thermodinamic Equilibrium (LTE) of plasmas.

1. Introduction

Technique Laser-Induced Breakdown Spectroscopy (LIBS) is a powerful method in spectroscopic by detection and quantification of chemical species in materials solid, liquid or gaseous phases, which analyzes the spectrum of radiation emitted by the breakdown or plasma generated by laser radiation. These plasmas-laser are a great source of spectral information from which we study different effects and physical processes that occur inside the plasma. In addition, they provide information of your response and behavior under the action of external electric and magnetic fields [1-3]. In this work the technique LIBS was employed for the study of alloys prepared in laboratory similar to mercury-based dental amalgams.

Mercury is a hazardous material for the health and the environment [4-6]. It is a heavy metal that at room temperature is liquid with a high density of 13.55 g/cm^3 at 25 °C. These properties of mercury liquid make difficult the obtaining of LIBS Spectra due to splash of liquid in the process of interaction of radiation with material, and other hand, with the increase of the temperature above 40 °C is generated toxic and corrosive vapor of mercury that requires careful handling. Failing that, due to its ease of alloyed with other metals, which melted at room temperature assume a crystalline form with certain properties, manufactured tablets with alloys composed of liquid mercury, Silver, Tin and Copper (amalgam), on which the laser plasmas were generated.

Because to its mechanical and thermal properties, amalgams are still used by dentists for filling and curing of dental caries. The amalgam is a metallic alloy grey composed of mercury liquid (50%) and matrix (50%), which is composed of silver (45%), copper (24%), Tin (31%), and sometimes zinc. The mercury reacts chemically with other metals creating a malleable paste that hardens with certain speed adhering in the dental cavity and facilitating the restoration of the damaged tooth. The mercury in dental amalgams is in its crystalline form that is little toxic. The mercury in dental amalgams is in its crystalline form is slightly toxic. However, practically the mercury evaporates at room temperature generating mercury vapour highly toxic to human health Therefore, it is possible that dental amalgams in teeth of people release certain amount of mercury vapor to consume drinks and food hot over 40 °C [7].

The plasmas-laser were formed on the solid matrix composed of Ag-Sn-Cu and on solidified amalgams mercury containing with different proportions or concentrations. The laser plasmas spectra were registered and analyzed in the spectral region of 250 nm to 900 nm. As expected, we identified atomic and ionic lines of the elements belonging to the matrix, whereas the amalgam spectra revealed additionally of mercury lines, without the presence of other impurities. The spectral analysis of the results shows on the one hand an excellent quality of purity of the commercial materials used for preparation of dental amalgam, and a good method of preparation samples. As regards the analysis of spectra, there is evidence of a clear increase in the intensity of characteristic lines of neutral mercury (Hg I) with the increase of the amount of mercury in different proportions employed (3:2, 5:2 and 6: 2); we observed also an intensification and a shift in some lines of Sn I in the spectra of the amalgam by effect of the addition of mercury, respect to the same lines in the spectra of the matrix without the presence of mercury. Based on the spectral measurements, and reported atomic parameters from characteristics lines of the elements used, was made estimation of the electron temperature of plasmas in conditions of Local Thermodynamic Equilibrium (LTE) and their behavior with the concentration of mercury [8].

2. Amalgam preparation and experimental description

The preparation of tablets of dental amalgam in the laboratory was an appropriate alloy with materials for dental purposes in healing or filling of dental caries, acquired in the commercial distributors, using a mechanical amalgamator. For the preparation of amalgam we used a powder mixture containing Ag, Sn and Cu which is introduced into the amalgamator together with liquid mercury for one minute. The soft paste (amalgam) is introduced rapidly into a plastic cylindrical tube that to harden takes the form of a small disc of variable diameter and thickness, which depends on the amount of materials used and the diameter of the cylindrical tube. The induced plasmas-laser (LIP) or plasmas-laser were generated on the samples always at the same focal position of the focusing lens of the laser radiation to guarantee the same excitation conditions and the formation of plasmas. Because to their chemical and structural properties liquid mercury is attached only on the surface of the solid, which is desirable as a restorative dental material and for purposes LIBS.

In this work, we implement methods of the LIBS technique [9] which can be applied for monitoring almost in real-time of mercury in dental amalgams. Plasmas-laser of the matrix Ag-Sn-Cu and of the amalgams were generated on the surface of samples focusing with a lens of 10 cm focal length the radiation of a pulsed Nd: YAG laser operating at 532 nm, at a maximum energy 80 mJ per pulse, duration of pulses of 3-5 ns and a frequency of 2 Hz. For the registration of spectra, the plasmas-laser radiation was conducted by an optical fiber grade UV-vis to the entrance slit of a optical spectrograph mounting Czerny-Turner of 0.25 m focal distance, equipped with a diffraction grating of 1200 grooves/mm, which produces a spectral resolution of 0.25 nm in the first order, and a CCD linear array of 2048 pixels. The spectra were processed and averaged with Spectral Array software with integration times for taking the spectra of 400 ms and 10 scans. In order to ensure the reproducibility of the plasmas-laser, the pellets were mounted on a positioning system based on stepper motors that allowed zoom out or zoom the sample of the focus point of the lens with a precision of 0.005 mm and/or rotate it with a resolution of 0.6 degress. As a source calibration of the system was used a spectral lamp Hg/Ar calibration that emits lines of 250 nm to 950 nm. During acquisition of spectra interference filters were used to avoid overlapping of spectra of superior orders and Noch filter at the entrance of the spectrograph to reduce the reflected or scattered laser radiation at 532 nm.

3. Results and discussion

Figure 1 shows a superposition of the spectra of the plasmas-laser of the matrix Ag-Sn-Cu (bottom) and the amalgam based on mercury (upper) in the region centered at 300.009 nm. In this region the spectra show atomic lines of neutral atoms Ag I, Sn I, Cu I, Hg I and the line 335.208 nm of Tin once ionized Sn II. In this sector also observed a shift towards long wavelengths of the Sn I lines in the spectrum of amalgam respect to the same line in the spectrum of the matrix. This fact we attribute to the presence of mercury atoms in the plasma formed, which come to play an important role in the process of interaction between the particles and electric fields.

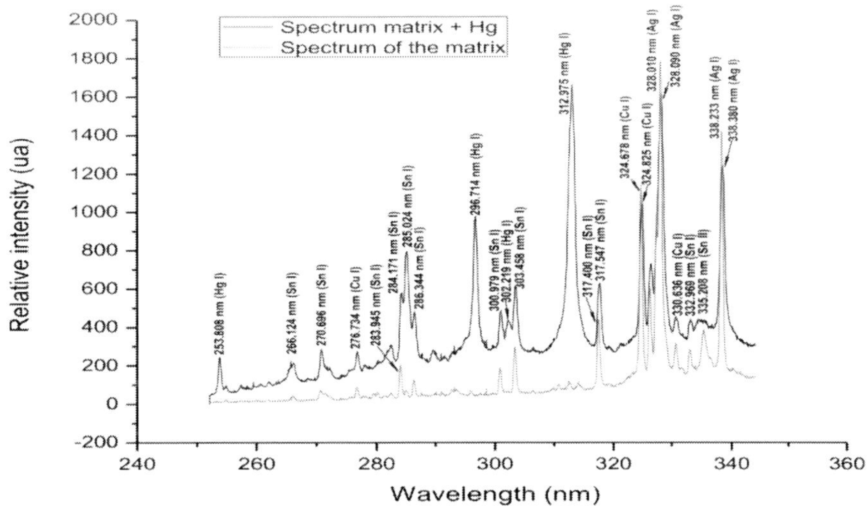

Figure 1. Spectra of plasmas- laser the matrix Ag-Cu-Sn (lower), and mercury-based dental amalgam (upper) in the spectral region centered at 300,009 nm.

The linear fit of the experimental points in figure 2 show the trend of the increase of the relative intensity of the characteristic lines 364.921 nm and 404.976 nm of Hg I, with increasing the amount of mercury employed in different proportions in the manufacture of amalgams (3:2, 5:2 and 6:2). With previously established calibration curves, the LIBS technique can be applied successfully by professionals in dentistry for the monitoring of the concentration of mercury in the manufacture or commercial dental amalgam [10].

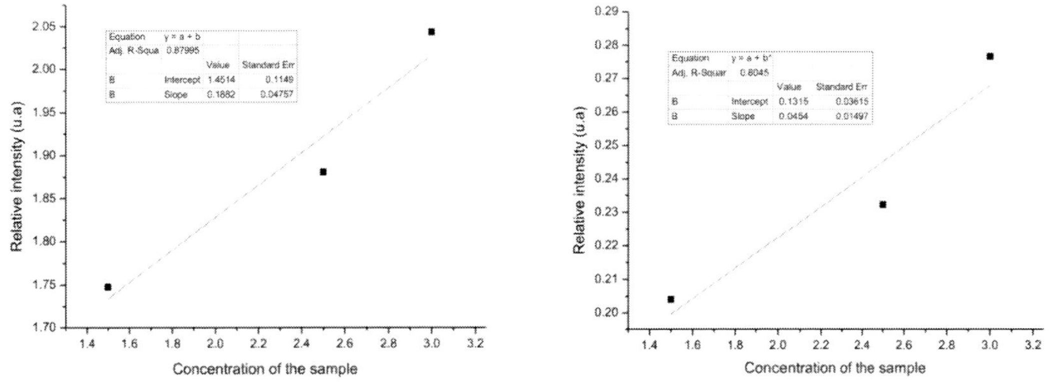

Figure 2. Graph relative intensity of lines 404.976 nm and 364.921 no of Hg I in term of the concentration of mercury in the amalgam.

Figure 3 shows the behavior of the electron temperature of plasmas of amalgams to the different proportions of Hg used (3:2, 5:2, 6:2). Each experimental point on the curve represents the electron temperature calculation using the ratio of relative intensities of the lines 364,921 nm and 404,976 nm of Hg I. The other atomic parameters: coefficients of probability, statistical weights, and energy of the upper levels of the transitions were obtained from the database [11]. The estimated values of the electron temperature for the respective proportions were 20 846 K, 19 139 K, and 16,872 K, using the Boltzmann population distribution of energy levels associated with spectral lines in conditions of Local Thermodynamic Equilibrium (LTE) of plasmas.

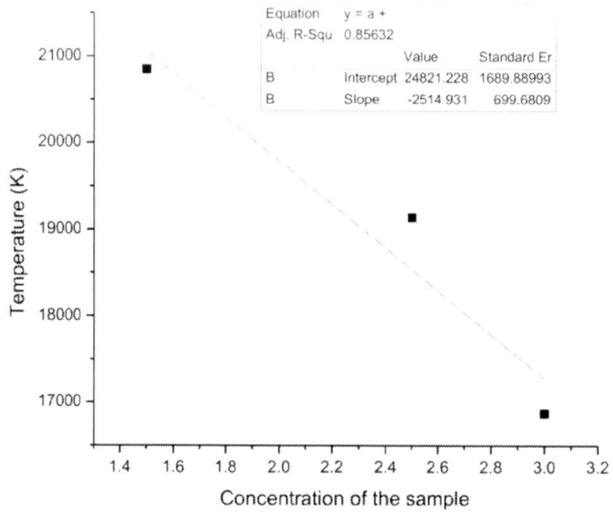

Figure 3. Behavior of the electron temperature of plasmas the amalgams in the proportions of mercury (3:2, 5:2 and 6:2).

4. Conclusions

The analysis of the spectra of the matrix Ag-Sn-Cu and the mercury-based amalgam, reveals that the commercial materials used for dental amalgams are of excellent quality and free from impurities which were not detected with the established LIBS technique. The spectral results show that with the LIBS technique can be able to monitor mercury in almost real time in dental amalgam employed by dentists in curing tooth and restoration of teeth.

References

[1] Eliezer S 2002 *The Interaction of High-Power Lasers with Plasmas* (Bristol and Philadelphia: institute of Physics Publishing)
[2] Cremers D and Radziemski L 2006 *Handbook of Laser-Induced Breakdown Spectroscopy* (England: John Wiley & Sons)
[3] Miziolek A, Palleschi V and Schechter I 2006 *Laser Induced Breakdown Spectroscopy:Fundamentals and Applications* (New York: Cambridge University Press)
[4] Foo S, Ngim C H, Salleh I, Jeyaratnam J, Boey K W 1993 *Environ Res.* **60** 267
[5] InSug O, Datar S, Koch C J, Shapiro I M, Shenker B J 1997 *Toxicology* **124** 211
[6] Sandborgh-Englund G, Elinder C G, Johanson G, Lind B, Skare I, Ekstrand J 1998 *Toxicol Appl Pharmacol* **150** 146
[7] Morales I y Reyes R 2003 *Rev Saúde Pública* **37** 266
[8] Thorne A, Litzén U, Johansson S 1999 *Spectrophysics: Principles and Applications* (Germany: Springer Verlag)
[9] Sierra L M, Rosellon L, Pacheco P, Álvarez J, Sarmiento R 2012 *Opt. Pura Apl.* **45** 263
[10] Tawfik W, Saafan A 2006 *International Journal of Pure and applied physics* **2** 195
[11] http://physics.nist.gov/PhysRefData/ASD/

Journal of Physics: Conference Series **466** (2013) 012018 doi:10.1088/1742-6596/466/1/012018

Breast phantom for mammary tissue characterization by near infrared spectroscopy

D A Miranda[1], K L Cristiano[1] and J C Gutiérrez[2]
[1] CMN, CIMBIOS, Universidad Industrial de Santander, Bucaramanga, Colombia.
[2] Grupo de Investigación en Química Industrial, Universidad Industrial de Santander, Bucaramanga, Colombia.

E - mail: dalemir@uis.edu.co

Abstract. Breast cancer is a disease associated to a high morbidity and mortality in the entire world. In the study of early detection of breast cancer the development of phantom is so important. In this research we fabricate a breast phantom using a ballistic gel with special modifications to simulate a normal and abnormal human breast. Optical properties of woman breast in the near infrared region were modelled with the phantom we developed. The developed phantom was evaluated with near infrared spectroscopy in order to study its relation with breast tissue. A good optical behaviour was achieved with the model fabricated.

1. Introduction

Breast cancer is associated to a high prevalence, morbidity and mortality [1]. The most effective method to minimize the prevalence, morbidity and mortality associated to the cancer is its early detection [2]. The early detection of breast cancer is an active research field [3, 4]. Near infrared (NIR) spectroscopy has been applied in the study of angiogenesis and it's has been suggested as a way to the early detection of breast cancer [5, 6, 7].

A special modification in abnormal breast tissue is the changes in concentrations of oxy and desoxy-hemoglobin associated to the angiogenesis [4, 8]. To monitoring the oxy- and desoxy-hemoglobin NIR spectroscopy has been suggested and proposed as a technique to the early detection of cancer [4] but in order to develop technological devices that applied this technique, physical models, i.e. phantom, that modelling normal and abnormal tissues are needed.

In this study we report phantoms that modelling the optical NIR properties of human breast in the wavelength range from 750 nm to 860nm. These phantoms were development in order to implement digital devices to the early detection of breast cancer.

2. Materials and methods

A gel based in the pectin poly-galacturonic, glycerin and water was prepared in the laboratory. Different proportion of the pectin, glycerine, water and ferric chloride were mixed in order to fabricate three different types of phantom, see figure 1. These phantom were molded and solidified in order to obtain an as human breast tissue model. Thin films (near to 1 mm) of phantoms were cut from the bulk material to measure the optical absorption coefficient and to compare with a reference data from [5].

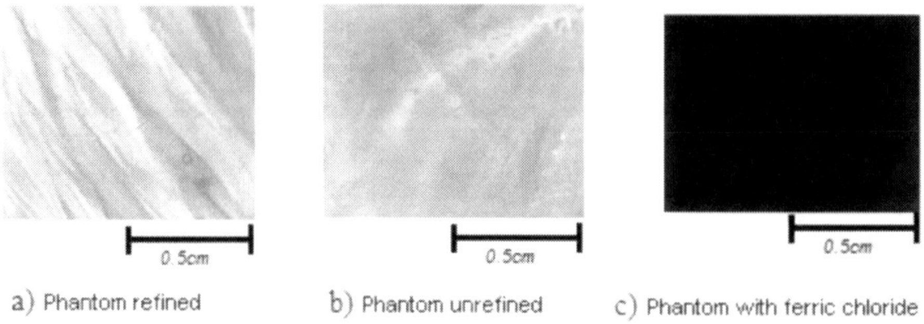

Figure 1. Three types of phantom fabricated with (a) pectin refined, (b) phantom with pectin not refined (with collagen) and (c) phantom with ferric chloride.

3. Results and Discussion

The fabricated phantoms were characterized by NIR spectroscopy. Optical absorbance of phantom (see figure 2) shows behaviour similar to the breast tissue [5]. Abnormal breast tissue is associated with angiogenesis [3, 8] and exhibit a higher optical absorption than normal tissue because the increasing of hemoglobin that increase the optical absorbance of breast. Then, the phantom with ferric chloride has an absorbance higher than normal breast tissue and could be associated with abnormal tissue. Phantom with refined pectin has an intermediate behaviour between normal and abnormal breast tissue. We suggest an increasing in the ferric chloride concentration could be used to model a more abnormal breast tissue. A mixing of the three different phantoms could be used to model different breast tissue state that could be applied to develop new technology to the early detection of breast cancer.

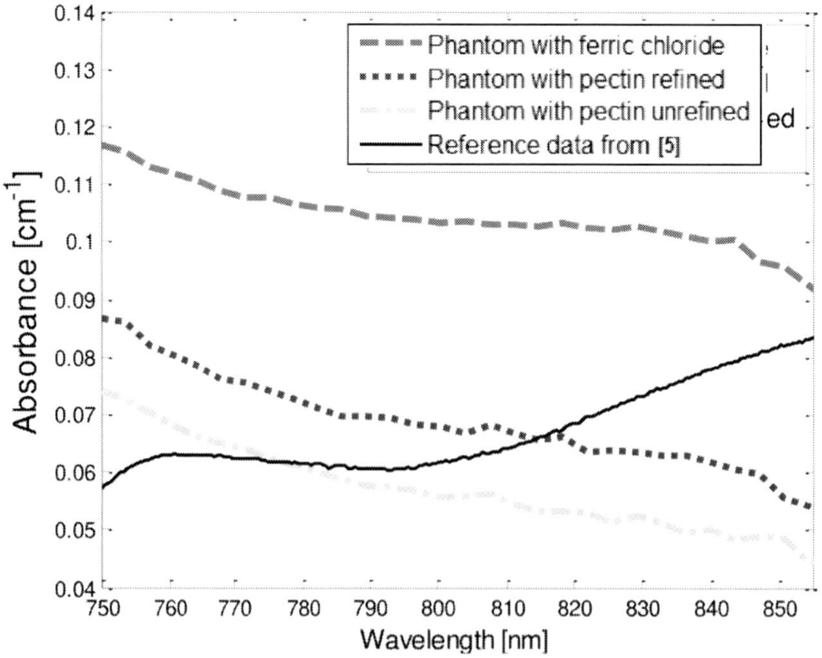

Figure 2. Absorbance of fabricated phantom compared with normal breast absorbance [5].

4. Conclusion

Three phantoms that model human breast tissue were fabricated and characterized by optical NIR spectroscopy. Behaviour similar to normal and abnormal tissue was observed in the absorbance of phantom. An advantage of fabricate phantom is that these are based in not toxic compounds, are biodegradable and recyclable, which is good for the environment.

Acknowledgements

We are grateful to the researchers of CIMBIOS and "Grupo de Investigación en Química Industrial" at the "*Universidad Industrial de Santander*" for his collaboration during the development of the experiments in special to Daniel Triana and lecturers: Samuel Jaimes, Janeth Fernandez, Sandra Pinto and Jose Luis Pinto for his collaboration in the development of the Phantoms and its characterization. We have special acknowledgements to the "Laboratorio de Análisis Instrumental" at the chemical school of the "*Universidad Industrial de Santander*" for the measure time.

References

[1] American Cancer Society 2009-2011 *Datos y Estadísticas sobre el Cáncer entre los Hispanos/Latinos* (Atlanta: Georgia http://www.cancer.org/)
[2] Herschel W Lawson, Rosemarie Henson, Janet Kay Bobo, Mary K Kaeser 2000 *Center for Chronic Disease CDC* **49-02** 35-55
[3] Smith R A, Saslow D, Sawyer K A, et al 2003 *C A Cancer J Clin* **53** 41-169
[4] Adhemar Longatto Filho, Jose Manuel Lopes, Fernando C Schmitt 2010 *Journal of Oncology* 576384
[5] Regine Choe 2005 *Diffuse optical tomography and spectroscopy of breast cancer and fetal brain* (United States: University of Pennsylvania)
[6] Erica Heffer, Vivian Pera 2004 *Journal of Biomedical Optics* **9-6** 1152–60
[7] J Todd Kuenstner, Karl H Norris 1994 *Journal Of Near Infrared Spectroscopy* (UK: NIRS Publications, ISSN 0967-0335)
[8] Henry C Pitot 2002 *Fundamentals of oncology* (New York: Marcel Dekker Inc Copyright)

Journal of Physics: Conference Series 466 (2013) 012019 doi:10.1088/1742-6596/466/1/012019

Optically induced birefringence in a azopolymer based on the phenomenon of photo-isomerization

E Mercado-Gutiérrez[1], S Valbuena-Duarte[2] and F Racedo-Niebles[1]
[1] Grupo de Espectroscopia Óptica de Emisión y Laser – GEOEL, Universidad del Atlántico, Barranquilla, Colombia.
[2] Grupo de Matemáticas e Informática de la Corporación Universitaria de la Costa MATINCUC, Barranquilla, Colombia.

E-mail: fran@mail.uniatlantico.edu.co

Abstract. The following article is a study of optically induced birefringence in an azopolymer Hema - HDR13. Generating anisotropy in the material through the photo-isomerization process for azo - compounds, chromophores are reoriented in the material through an external electric field, from an Nd: YAG laser, emitting with a wavelength of 532.8nm Based on an absorption spectrum on an azopolímero samples was determined that the wavelength at which the phenomenon is given. Measurements were obtained for samples birefringence how efficient are able to observe the samples so that they can be used as raw material for optical storage devices. Varying the excitation laser power (Nd: YAG) were observed variations in signal strength, making observe a maximum saturation power chromophores which are reoriented in the sample. Also analyzed the theoretical models of photo - molecular orientation, and succeeded in identifying which of them corresponds to the results and establishing the characteristic times appearing in each for Hema - DR13.

1. Introduction

The azopolymers are compounds which react to a stimulus characteristic radiation, causing a change in some physical and chemical properties in them. Some physical properties that may occur in the molecule is the change in the electric dipole moment, and change in geometry due to an inversion or rotation mechanism. This is thanks to the feature that shows the azo, azo functional group-N = N-azo. The used in the development of this work was the 4 - [N-ethhyl-N-(2-ethyl-methacryloxy)] -amino-2-chloro-4-nitroazobenzene, known for short as HEMA - DR13 which is a compound that has a very stable and has been the subject of many studies and research based on this phenomenon. Figure 1 shows the chemical structure of HEMA DR13, where one can observe the azo functional group attached to two aromatic rings which has a large carbon chain, with the particularity that the opposite ends of the molecule have a high electronegativity together. The functionality of this feature consists in the provision of the molecule to behave like an electric dipole in presence of an external electric field [1].

The photo isomerization process occurs when the linearly polarized light of a specific wavelength located in the range of absorption of the sample is incident on the molecule. In the steady state the molecule is in the trans isomer which is the lowest energy when the molecule is excited, and changes the cis isomer by means of the rotation mechanism or the inversion of the aromatic rings.

The photo - isomerization process trasn - cis - trans anisotropy occurs in the molecule giving rise to the photo-induced birefringence [2].

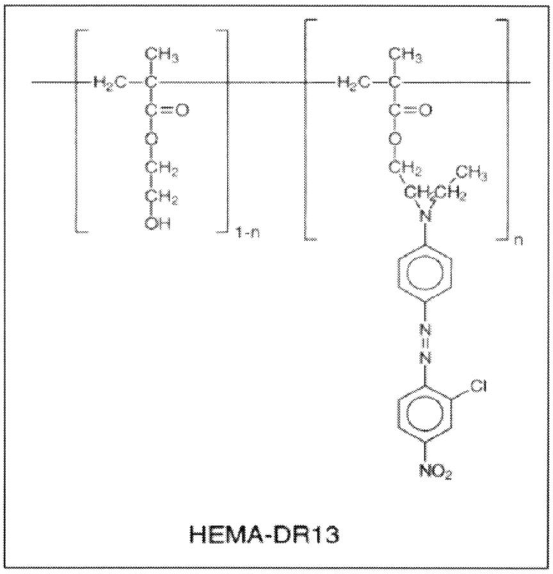

Figure 1. Chemical structure of HEMA - DR13.

Using two lasers, laser He - Ne linearly polarized low power with a wavelength of 632.8nm, and a Nd: YAG laser with a wavelength of 532.8nm and also linearly polarized low power are able to induce the optically birefringence in a microstructured films HEMA - DR13, which were prepared by the deposition method in a saturated atmosphere. Using the optical arrangement shown in figure 2 is able to induce birefringence. The reading laser (He - Ne) is in charge of verifying the state of polarization in which the chromophores are azopolimero through two polarizers which are crossed to each other. When the excitation laser (Nd: YAG) impinges on the sample starts the process, allowing the external electric field provided by the laser passing chromophores excited state so trasn - cis. To make this happen successfully, be careful that the polarization direction of the laser beam is perpendicular to the reading provided by the laser beam writing. As time passes, the chromophore in the sample are aligned parallel to the external electric field of the laser writing [3, 4].

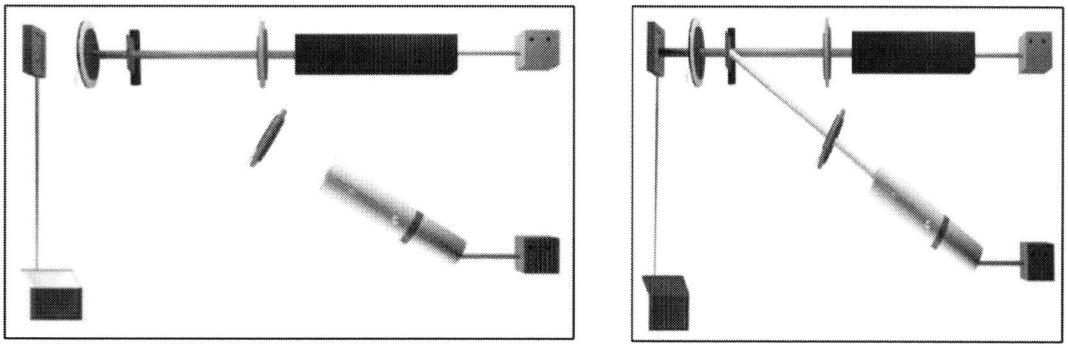

Figure 2. Schematic experimental photo - induced birefringence before and after exciting the sample.

2. Results and discussion

To establish the wavelength of the lasers used, it was necessary to take an absorption spectrum as needed that one of the lasers (the reading) will not affect the state of the chromosphores in the sample, but it was totally "transparent" to it. The result obtained is shown in figure 3, which shows that the spectral range in which the sample absorbs more light corresponds ranging from 520 to 600nm, a range in which is the wavelength of the excitation laser. While the laser reading is outside this range, therefore the specimen is unaffected by this.

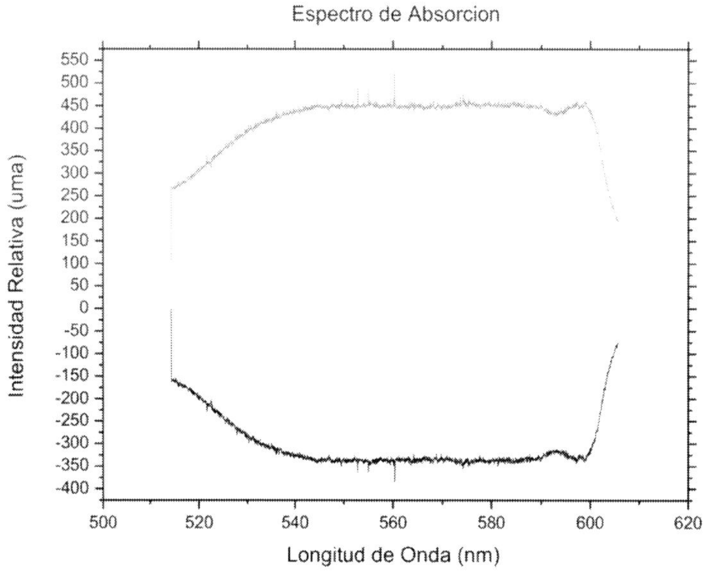

Figure 3. Absorption spectrum of HEMA - DR13, the spectral window in which the phenomenon of photo-isomerization.

The signal obtained was taken with a photodetector with a rise time of 1ns, and Labview a data acquisition card, processed through a pc. The signal obtained for the sample shown in figure 4. It shows the birefringence induced in the sample azopolymer for a 4.5mW power. Clearly shows the rapid increase of the signal, this is the time when the writing laser is incident on the sample, ensuring that the chromophores will begin to move in the polymer layer.

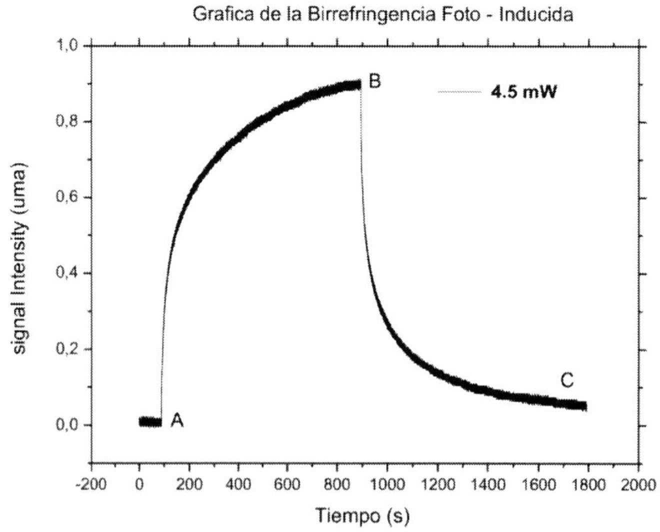

Figure 4. Intensity of the signal obtained for the photo-induced birefringence versus time.

After the same process was performed but varying the write laser power, to obtain the results shown in figure 5.

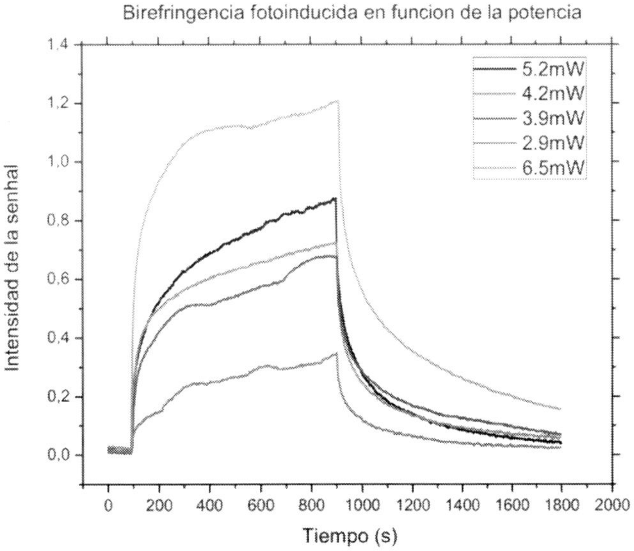

Figure 5. Induced birefringence signal depending on the power.

Clearly shows that as the power increases the signal intensity is increased to approach a maximum value. At this point most of the chromophores present in the sample are in the excited state, the cis state. Immediately remove the chromophores laser writing start state to unwind from cis - trans, showing that the signal strength decreases gradually to its initial value. This process is repeated for each of the samples used. The temperature at which the samples were in the laboratory was 24 °C. the signal strength varies according to the temperature showing that the lower the temperature the higher the birefringence signals.

3. Conclusions

It was possible to obtain an absorption spectrum of the samples prepared by the method of deposition in a saturated atmosphere was achieved which identify the range of absorption of the sample to the utilization of lasers both read and write. Furthermore, the signals obtained for a fixed power birefringence achieving verify and observe the photoisomerisation in the HEMA - DR13 achieving a relaxation time of about half an hour with the quality of this material can be used for optical storage utilization. Furthermore it could be shown that for the variation of the excitation power of the signal intensity variation, noting that between higher the excitation power, the higher the intensity of the photo-induced birefringence.

References

[1] E Mercado Gutierrez, A Quiroga Soto, F. Racedo-Niebles 1992 *J. OPA.* B **43-3** 303
[2] Maria Rigaborda, Estibaliz Merino 2009 *An.Quim*, **105-4** 290
[3] Weininger S and Stermitz F 1998 *Qumica Organica (*Barcelona: Reverté)
[4] M C Cardoso 2005 *Estudo da Birrefringencia Fotoinduzida Por um e Dois Fotonsem Compostos Azo-aromaticos da Familia Salen* (São Paulo: Universidad de Sao Paulo)

Journal of Physics: Conference Series **466** (2013) 012020 doi:10.1088/1742-6596/466/1/012020

Conducting polymers doped with a mineral phase: structural and electrical study

C P González[1], A M Montaño[1,2], S Estrada[1] and C Ortiz[3]
[1] Grupo de Investigación en Materiales - GIM, Departamento de Ciencias Básicas, Universidad Pontifica Bolivariana Seccional Bucaramanga, Floridablanca, Colombia.
[2] Escuela de Química, Universidad Industrial de Santander, Bucaramanga, Colombia.
[3] Grupo de Investigación en Superficies, Electroquímicas y Corrosión - GSEC, Universidad Pedagógica y Tecnológica de Colombia, Tunja, Colombia.

E-mail: claudia.gonzalez@upb.edu.co

Abstract. This work reports the results obtained of a series of novel doped conducting polymers (CPs) of polyaniline/hematite (PANI/HEM), which were synthesized in acidic aqueous solution by the in situ chemical oxidative polymerization, using ammonium peroxydisulfate as oxidant reagent. The synthesis was carried out with 20, 40 y 60 % (weight percent) contents of hematite (HEM) at 8 and 14 h of polymerization times (tP). These composites were structurally characterized by X-ray diffraction (XRD) and Fourier-transform infrared spectroscopy (FTIR). An electrochemical analysis was made by Electrochemical impedance spectroscopy (EIS). Results of this study allow to evaluate the influence of hematite on the improvement of the structural properties and in the increase of the electric conductivity (σac) of the doped polymers compared to CPs without dopant agents.

1. Introduction

Nowadays conducting polymers (CPs) are an interesting research area because of these several applications as low impact environmental devices with longer half life time. It has been used in batteries manufacturing, muscles and artificial nervous, intelligent optical (windows, mirrors) parts, electromechanical devices and as additives against corrosion process, between many others [1-5]. Besides it has been proved that CPs improves its optical, mechanical and electrical properties by addition of doping agents as fly ashes [6-8]. The goal of this study is to evaluate the effect on structural and electrical properties of emeraldine, Poly-aniline (PANI), a widely used CP, by doping it with hematite mineral (HEM).

2. Experimental

PANI/HEM composites were synthesized through oxidative polymerization adding ammonium peroxydisulfate in acidic media with doping agent *in situ*. HEM was added at 20, 40 and 60% in weight. Mix reactions were keeping during 8 and 14 hours as polymerization time. Samples of doped PANI were characterized by XRD, FTIR and EIS.

3. Results and discussions

In all PANI/HEM diffractograms the doping achieved is seen figure 1. By comparing diffraction patrons it is also possible to affirm that during polymerization there are not structural changes in polymer or in dopant mineral, as PANI/HEM does not show new reflections compared with precursors. Doping could be seen as dopant dispersion at polymer matrix. This dispersion produces an increase in crystal degree of doped CPs. Reached doping is also affected by polymerization time. Diffraction patron of the PANI/HEM with same HEM concentration, it has some variations at intensities at 8 and 14 h of t_P for both, PANI and HEM signals. In figure 2, intensity for peak at 25,35° of PANI/HEM as function of HEM concentration for both t_P is shown. By increasing of dopant mineral, intensity of same peaks for PANI/HEM has trend to decrease, however, this tendency is lightly major when polymerization reaction is carried out at 14 h. This fact could mean that at this t_P a higher doping is produced and therefore, a greater degree of crystallinity.

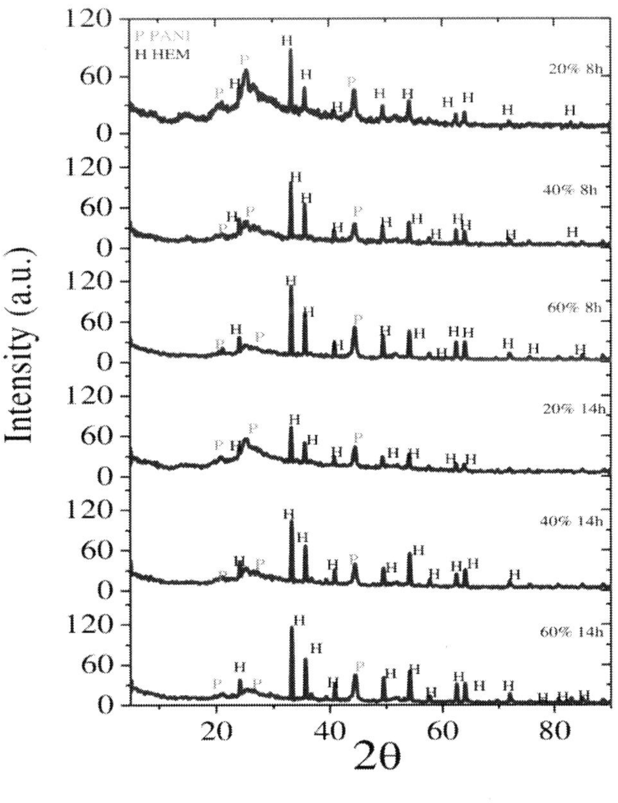

Figure 1. X- ray diffraction patterns for PAni/Hem polymers doped.

FTIR spectra for PANI, HEM and PANI/HEM composites are shown in figure 3. PANI FTIR spectra shows its eight characteristics bands. At 3416 cm^{-1} it is seen tension vibrations of N-H of amine groups, at 2923 cm^{-1} is due to C-H stretching vibration of aromatic rings. Signal attributed to quinolinic rings presented on the oxidation state of emeraldine salt (PANI), is at 1574 cm^{-1} .while the band close to 1482 cm^{-1} is assigned to C=C stretching vibration of aromatic ring. The doublet at 1298 y 1243 cm^{-1} corresponds to stretching vibration of C-N bond of secondary amine. Band at 1139 cm^{-1} is the flexion vibrations on aromatic plane for C-H, while deformations C-H out of plane of a 1,4-disustitued system it appears at 799 cm^{-1}.

HEM FTIR presents seven characteristics bands. At 3411 y 1629 cm^{-1} are bands of stretching and bending vibration of water of HEM hydration. Band at 1220 cm^{-1} is due to Si-O-Si bonds vibrations of quartz. At 1138, 898 y 798 cm^{-1} there are bands corresponding to the stretching metal–OH bond vibrations for the iron and silice compounds like SiOH, FeOH and goethita (α-FeOOH). Finally, the most intense bands for HEM at 551 y 462 cm^{-1} are keys to identify doped PCs.

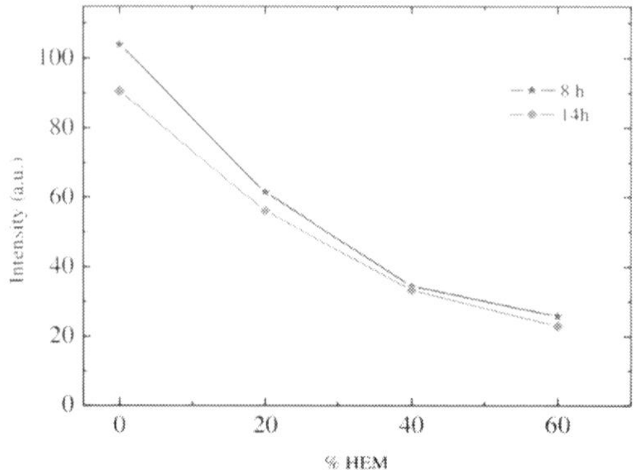

Figure 2. PANI peak (25, 35°) intensity against HEM percentage at 8 y 14 h tP. (Solid lines are a visual guide to the trend).

FTIR spectra for PANI/HEM composites show every single PANI bands. About HEM, in these spectra it is possible to identify bands at 552 y 480 cm^{-1} at 8 h t_P and 568 y 461 cm^{-1} at t_P of 14 h, intensity of all signals increases with increasing HEM concentration on polymer matrix. Rising dopant percentage results in a growing of almost all bands at right end of the spectrum (for both PANI and HEM) in relation to the broad band at 3416 cm^{-1} corresponding to N-H bond of PANI. In general, HEM bands at 14 h are most intense that those at 8h t_P. It is seen for example, that with HEM 20%, mineral characteristics bands are almost imperceptible at 8h t_P whereas at 14 h are visible clearly.

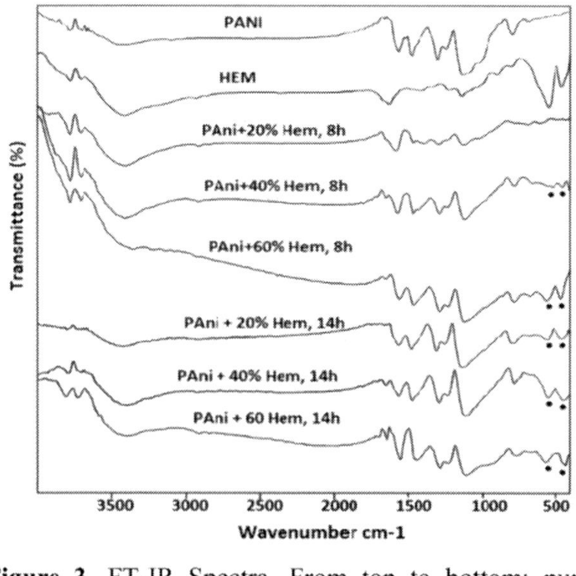

Figure 3. FT-IR Spectra. From top to bottom: pure PANI, pure HEM, doped CPs 8 h tP, doped CPs 14 h tP.

Based on the conductivity measurements it is possible to establish that PANI/HEM composite at 20% HEM with 8h t_P has the higher σ_{ac}, its value is four times the conductivity of PANI without doping. There is an optimal dopant concentration with which the maximum σ_{ac} for doped CPs is achieved (20% HEM), an increase in dopant content above this value has a contrary effect on conductivity, from this point σ_{ac} diminish to values even lower than those for pure PANI. It behavior for both reaction times is schematized in figure 4.

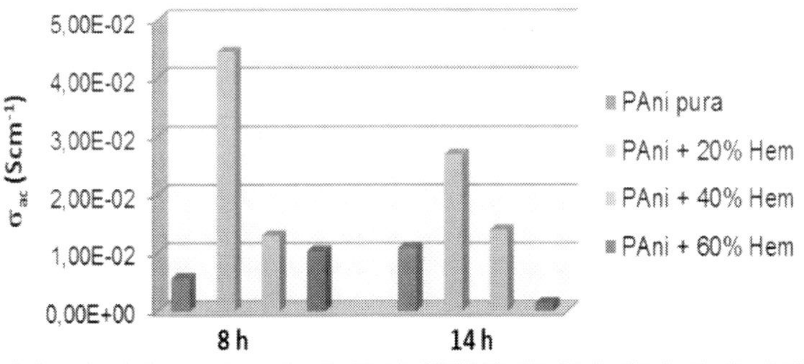

Figure 4. σ_{ac} values PANI/HEM against HEM percentage (t_P = 8 y 14 h; $f = 1x10^5$ Hz).

4. Conclusions

Experimental techniques XRD and FTIR allowed to confirm the presence of HEM on doped CPs and also to follow doping process by changes in dopant concentration and reaction time. It was found that, the rise in CPs doping degree and crystallinity is proportional to the concentration increase of doping agent, being higher in PANI/HEM compounds synthesized at 14h t_P. It was determined that the biggest increment of Cps conductivity is reach at 20% of dopant agent, what can be said that there is an optimal concentration of HEM which leads to a maximum value of conductivity. Any additional amount of HEM produces a negative effect on PANI/HEM conductivity.

References

[1] Narayan H, Alemu H and Iwuoha E 2006 *Physica Status Solidi* **203-15** 3665
[2] Lange U, Roznyatovskaya N and Mirsky V 2008 *Anal. Chim. Acta.* **614** 1
[3] Liu P 2008 *Curr Opin Solid State Mater Sci.* **12** 9
[4] Armelin E, Pla R, Liesa F, Ramis X, Iribarren J and Alemán C 2008 *Corros. Sci.* **50** 721
[5] Xia L, Wei Z and Wan M 2010 *J. Colloid Interface Sci.* **341** 1
[6] Bhadra S, Khastgir D, Singha N and Hee Lee J 2009 *Prog Polym Sci.* **34** 783
[7] H Narayan, H Alemu, V Somerset, E. Iwuoha, M Hernández, J Hernández, A Montaño and J Henao 2009 *Fly Ash: Reuse, Environmental Problems and Related Issues* (New York: Nova Science Publishers Inc.)
[8] H Narayan, Angela M Montaño, Monica L Hernández, July A Hernández, Claudia P González and César A Ortiz 2012 *J. Mater. Environ. Sci.* **3-1** 137

Journal of Physics: Conference Series **466** (2013) 012021 doi:10.1088/1742-6596/466/1/012021

Single crystal substrates effect on the critical behavior in $La_{2/3}Ca_{1/3}MnO_3$ thin films

J Pino[1], O Arnache[1], J Osorio[1] and L Tirado[2]
[1] Instituto de Física, Universidad de Antioquia, A.A. 1226, Medellín, Colombia.
[2] Grupo de Optoelectrónica, Universidad del Quindío, Armenia, Colombia.

E-mail: julipino@gmail.com

Abstract. In this work we study the substrate influence in magnetic properties and the critical behavior of $La_{2/3}Ca_{1/3}MnO_3$ thin films grown on single-crystal substrate of $LaAlO_3$, $SrTiO_3$ and $NdGaO_3$, using a DC-magnetron sputtering system. All films were growth under the same conditions and with thicknesses close to 130nm. We made measures of magnetization as a function of temperature between 150 and 300 K, around critical temperature (Tc) with different external applied magnetic fields (100 Oe – 30 kOe). X rays measures were made to verify the existence of $La_{2/3}Ca_{1/3}MnO_3$ (LCMO) phase, where we observed two peaks corresponding to the samples with Miller index (020) and (040) that verify the presence of LCMO phase in these thin films. Furthermore, critical exponent (β) was obtained fitting the magnetic measures with a proposed model. β values for each substrate are around 0.25, 0.27 y 0.36, that show an important influence of substrate.

1. Introduction

Manganites have been matter of study in recent years because they have presented the phenomenon of colossal magnetoresistance (CMR) which is promising for future technological applications, in particular the Manganites family of type La1-xAxMnO3 (A = Ca, Sr, Ba, Pb) [1]. This paper presents a study of critical magnetic behavior of the magnetization as a function of temperature in the vicinity of the magnetic transition zone for $La_{2/3}Ca_{1/3}MnO_3$ (LCMO) thin films, grown on single crystalline substrates of $LaAlO_3$ (LAO), $SrTiO_3$ (STO) and $NdGaO_3$ (NGO). The analysis of how the type of substrate influence on the critical exponents is done by characterizing the critical behavior (magnetic transition Ferro-Paramagnetic) of this compound.

2. Experimental procedure

LCMO thin films were grown on $LaAlO_3$ (LAO), $SrTiO_3$ (STO) and $NdGaO_3$ (NGO) single-crystalline substrates (100) under the same conditions deposition by using DC-magnetron sputtering system. During the deposition the substrate temperature was kept at 850 °C, the pressure of high purity oxygen was 500 mTorr, the substrate-target distance was 40 mm and the deposition power was 35 W. The X-ray diffraction measurements were performed with a Bruker D8 Advance, with a step of 0.02° and 3 s per step. Magnetic measurements were performed via vibrating-sample magnetometer technique (VSM) with the physical properties measurement system (PPMS, Quantum Design). All measurements were taken by using a 40-Hz vibration frequency for the detection coil with 2-mm The magnetization vs. temperature M vs. T after zero field cooling (ZFC) was measured with an applied external magnetic field H of 300 Oe to 30 kOe.

3. Results and discussion

After obtaining the films became the structural analysis by X-ray measurements. Figure 1 shows X-ray diffraction patterns associated with LCMO/LAO, LCMO/STO and LCMO/NGO thin films. All films are single phase without any detectable impurity or secondary phase. Figure 1 (a), shows that the most intense peaks correspond to substrates; the reflections in 23.0° and 47.06° with Miller indice (020) and (040), respectively, correspond to LCMO phase grown on LAO and STO, respectively. However, films grown on NGO present only one peak that belongs to LCMO associated to Miller indice (040) due to orthorhombic crystalline structure of both, by that reason the sample peaks are shielded by substrate peaks. All LCMO peaks on different substrates were fitted using Gaussian and Lorentzian functions. In the figure 1 (b), we show how looks the fit for one film grown on LAO. The lattice parameters were calculated with the peaks positions found from the fits. The values were a =5.4215(3) Å, b =5.4325(2) Å and c = 7.6855 (3) Å.

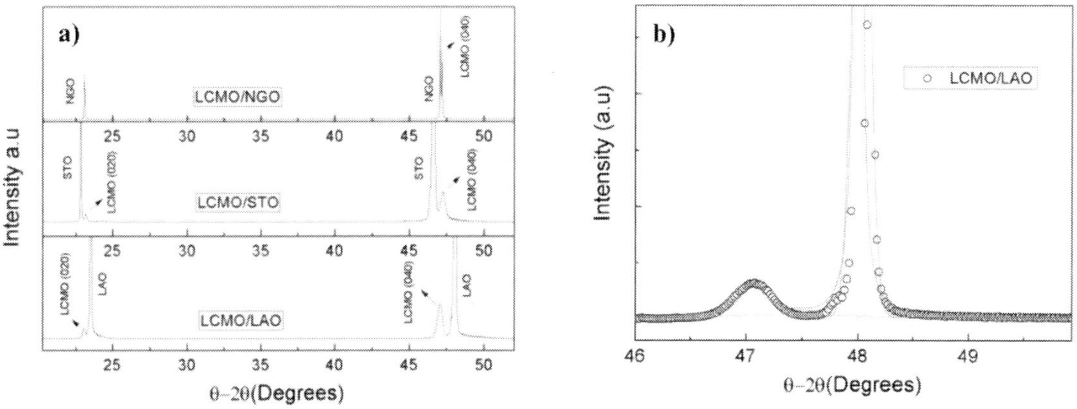

Figure 1. (a) XRD patterns of LCMO thin films on LAO, STO and NGO. (b) Open circles are experimental data and the continuous lines are the fits on LCMO/LAO.

Regarding the magnetic properties, in the figure 2 are shown the M vs. T curves taken in each of the samples. LCMO/LAO and LCMO/STO measurements exhibit the typical magnetic behavior of this compound. The change in the magnetic transition is not abrupt, i.e, magnetization is not zero in a precise value of temperature as is expected for an ideal ferromagnetic material. In this case there is a range of temperature where magnetization is becoming zero, this is attributed to the magnetic inhomogeneous nature of the samples. Such behavior is more evident with high magnetic fields. Furthermore, magnetization values are strongly influenced by the kind of using substrate. Films grown on STO have bigger magnetization values than those grown on LAO due to the coupling between the structure and each substrate. On the other hand, in films grown on NGO submitted to high magnetic fields (between H= 6 and 30 kOe), shows a different behavior of the others, the magnetic transition slope decreases as the H increases until the transition disappears. However, typical magnetic transition of LCMO maganites is observed at lower fields of 1000 kOe (see inset figure 2. (c)).

In order to analyze the critic nature of magnetic moment as a function of temperature around the transition region, the proposed model in [2] was used. Figure 2 (d), shows the fits obtained for each sample with external magnetic field of 1000 Oe, which are in excellent agreement with experimental data. Figure 3 (a), exhibit values of critical exponent β vs H given by the fit. For samples grown on LAO and STO, β varies in a non-monotonically for smaller field, with values between 0.14 and 0.25. Whereas that, for large fields > 4000 Oe, β values do not change considerably, these fluctuate between 0.24 and 0.27.This indicates that the critical magnetic behavior of the samples does not depend strongly on the substrates in this range of magnetic fields. However, in NGO samples the critical exponent β takes values between 0.28 and 0.36, (see inset figure 3 (a)) which are greater than the substrates previous, the highest value is related with magnetic behavior type Heisenberg [3].

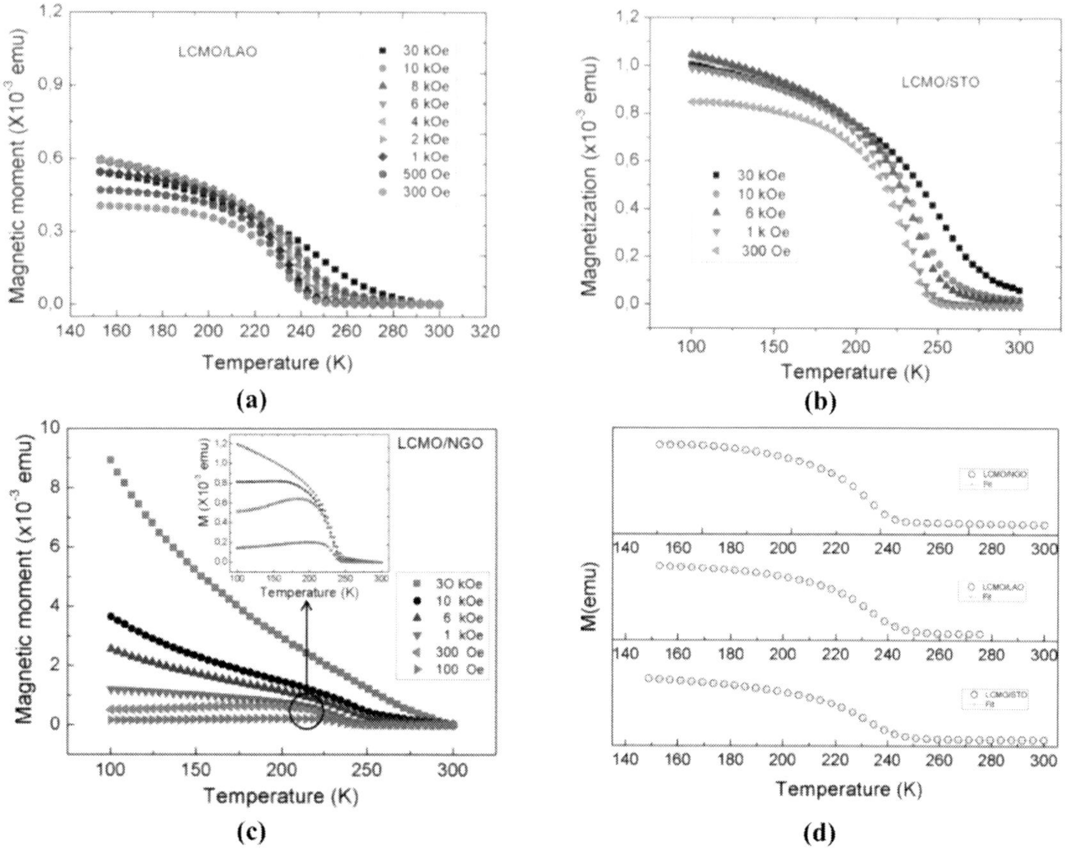

Figure 2. (a), (b), (c) Magnetic moment as a function of temperature. (d) Fit, circles are the experimental data and the solid line is the fit.

Extrapolating the β values from figure 3 (a) y (b), we found the β (H=0), these values are shown in table 1. LCMO/STO and LCMO/LAO films have β values of 0.27 and 0.26 respectively, these values are not related to a particular kind of universality, however we consider this result interesting because these values are typical of systems that exhibiting "tricritical" point values, which have been found in these same manganites systems [4]. As to magnetic transition width (ΔT) as a function of H were also obtained from the fits. As shown in figure 3 (b), we divide the ΔT behavior in regions (1) and (2). In region (1) ΔT does not vary with H for each of the substrates. Nevertheless in region (2) for H>4000 Oe, ΔT values tend to increase linearly with the applied field, in this case ΔT_c has contributions from both the magnetic inhomogeneity and the applied magnetic field. It is important to note that ΔT follows a power law type behavior as a function of H of the form $\Delta T = \Delta T_c + cH^{1/\eta}$ [2], where η is a critical exponent associated with the dependence of the magnetization with H around the T_C. ΔT_c is the width of transition to $H = 0$ and c is a constant dependent of the sample. η values were obtained from ΔT (H) fit using the above expression which are shown in table 1. For LCMO/LAO samples these values are nearly to Heisenberg type behavior, but for STO and NGO the found values are lower. Assuming the relation taken from [2] between the exponents $\eta = \beta.\delta$ values were calculated for each substrate which are between 3 and 7.

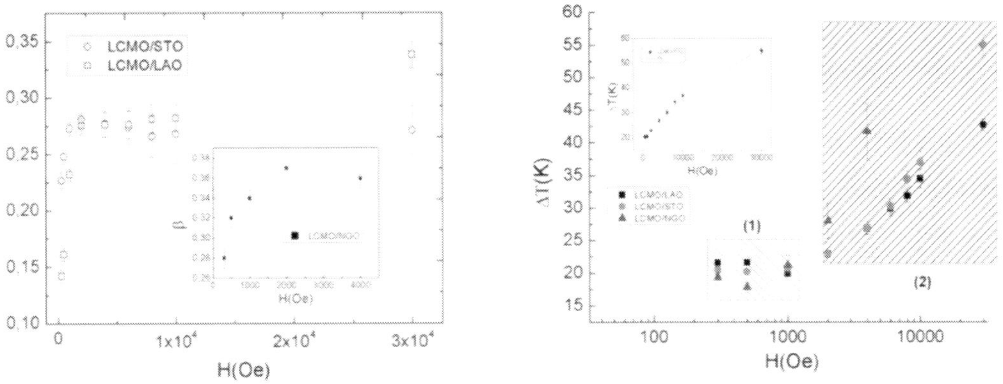

(a) **(b)**

Figure 3. (a) β exponent as a function of applied external magnetic field in LCMO thin films, (b) ΔT_C as a function of applied external magnetic field in LCMO thin films, the zone (1) and (2) are visual guides. The inset shows Polynomial fit on ΔT values for the LCMO/STO film. Squares are experimental data and line is the fit.

Table 1. Values for β (H=0), η and δ.

Substrate	β (H=0)	η	δ
$SrTiO_3$ (STO)	0.27 ± 0.01	1.34 ± 0.12	5.0 ± 0.6
$LaAlO_3$ (LAO)	0.26 ± 0.01	1.72 ± 0.20	6.6 ± 0.8
$NdGaO_3$ (NGO)	0.36 ± 0.01	1.21 ± 0.40	3.4 ± 1.2

4. Conclusions

In summary, in this work LCMO films were grown on STO, LAO and NGO substrates. The magnetic measurements show a large dependence with the substrate used. Different behavior was observed in the analysis done in LCMO/NGO films. β values obtained were 0.32 and 0.36 indicating that the magnetic behavior of this system is related to the Heisenberg model 3-D.

Acknowledgements
This work was supported by CODI – University of Antioquia under Project SIU 241434.

References

[1] Yoshinori T 2000 *Colosal Magnetoresistive Oxide* (New York: Science Pub.)
[2] Berger A, Campillo G and Vivas P 2002 *J. Appl. Phys* **91** 10
[3] Stanley H E 1993 *Introductions to phase transitions and critical phenomena* (New York: Oxford University Press)
[4] Kim D, Revaz B, Zink L and Hellman F 2002 *Phys. Rev. Letters* **89** 22

Analysis of magnetic and structural properties in $La_{0.6}Sr_{0.4}MnO_3$ ferromagnetic particles under the influence of mechanical ball milling effect

G Campillo[1], A Gil[2], O Arnache[2], J J Beltrán[2], J Osorio[2] and G Sierra[3]
[1] Departamento de Ciencias Básicas, Grupo de Materiales Nanoestructurados y Biomodelación, Universidad de Medellín, Colombia.
[2] Instituto de Física, Grupo de Estado Sólido, Universidad de Antioquia, Colombia.
[3] Facultad de Minas, Universidad Nacional de Colombia-Sede Medellín, Colombia

E-mail: gecampillo@udem.edu.co

Abstract. We have investigated the magnetic, structural and morphological properties of La0.6Sr0.4MnO3 (LSMO-40) manganite particles, synthesized by solid state reaction method. The resulting LSMO-40 powders were milled in air atmosphere during 3, 6 and 12 hours, by using a planetary ball milling. Samples obtained were characterized by X-ray diffraction, scanning electron microscopy - SEM and magnetization measurements as a function of temperature and magnetic field. A Rietveld analysis was carried on each XRD pattern, and was observed a reduction in crystallite average size (Dv) with increased ball milling time, tM. This is associated with a decrease in particle size. A characteristic rhombohedral crystal structure for the LSMO-40 phase was identified (space group R3C), independent of the milling time of the powders. However, from SEM microstructure was observed more homogeneity in the grain distribution by milling process. The results of magnetic characterization, showed that samples with higher tM (smaller grain size), presented the lowest value of the saturation magnetization, which is attributed to surface effects that induce magnetically disordered states with decreasing particle sizes. This magnetic anisotropy surface is evidenced also on the changes of coercive fields, HC, measured at low temperatures, which increased with increasing tM.

1. Introduction

Since the discovery of Colossal Magnetoresistance (CMR) phenomenon, which is caused by double exchange mechanism DE [1], in $La_{1-x}A_xMnO_3$ (A= Sr, Ca, Ba, etc) named manganites, it has been found new and exciting properties for potential applications in different areas [2, 3]. These doped manganites, which possess a perovskite structure, show a variety of phases depending of stoichiometric structure [4]. Particularly, $La_{1-x}Sr_xMnO_3$ (LSMO) manganite has been focus of important attention due besides of CMR effect, to its high Curie temperature (Tc), conduction by spin-polarized and close lattice parameter with others compounds, in a range of x concentration [5]. The physical properties of this systems are sensitive to size effect depending of the fabrication, in single crystal, thin films (lattice strain, interfaces effects) or powders (core-shell effects) [6-8]. In the case of particles at nanosize scale of LSMO, recently they have been much studied because of grain boundaries effects, which have shown significant differences in their physical properties respect to the bulk material, having very important applications. A reduction of grain size was found to influence the magnetization and coercivity in these compounds. Among the models proposed for describe these particular behaviours, has been found that the structural disorder near to grain boundary, affects the DE mechanism due to its sensitivity to changes in the stoichiometric composition, vacancies, dislocations, etc. This produces magnetically disorder states, in the surface of the grains. In this work we report a structural and magnetic analysis of $La_{0.6}Sr_{0.4}MnO_3$ (LSMO-40) powders, obtained by solid state reaction method, and subjected to mechanical ball milling. Models based on the contribution of the grain surface on the magnetization and coercivity, and the dependence of the magnetic properties with the particle size, is discussed.

2. Experimental

In this work, a pure phase of $La_{0.6}Sr_{0.4}MnO_3$ (LSMO-40) ferromagnetic manganite was obtained by solid state reaction method. The phase was obtained from the mixture of high purity precursors, synthesized according to the appropriated stoichiometric molar amounts. In order to reduce the particle size of the LSMO-40 sample, the powder was subjected to mechanical milling process in air atmosphere, by using a planetary ball milling Fritsh PULVERISETTE 5. For this study, were established three different milling times: 3, 6 and 12 hours, with a ball-to-powder ratio of 20:1 (using stainless-steel balls), for counting with a set of 4 samples (including unground sample – Non-Milling (NM)) of this manganite. The crystallographic structure of the powders was performed by X ray diffraction (XRD), with a Cu-Kα radiation source. All XRD patterns were fitted by using the Rietveld refinement analysis in order to obtain structural parameters, including the average crystallite size for each milling time, t_M. Finally, magnetic measurements were performed via vibrating-sample magnetometer (VSM) with the physical properties measurement system (PPMS, Quantum Design). All measurements were taken by using a 40-Hz vibration frequency for the detection coil with 2-mm. Magnetization curves as a function of the temperature M (T), were measured varying temperature from 5 to 375 K, for both Field Cooling (FC) and Zero Field Cooling (ZFC), with an external applied magnetic field of 500 Oe. Magnetization hysteresis loops M (H), were measured in the ranges of applied magnetic field from -5 kOe to 5 kOe.

3. Results and discussion

The X-ray diffraction is used to verify the crystal structure of the samples. In figure 1 (a) shows the XRD diffraction data and the Rietveld refinement, for all milling time of LSMO-40 samples. It is observed a good correspondence between the experimental data and the fit, which showed a single crystallographic phase, corresponding to the rhombohedral structure that belongs to space group R3C. This structural phase, characteristic for this manganite [9], is present for all t_M, including the sample without grinding, indicating that the process of synthesis for this concentration crystallizes in a single phase, which is stable under the grinding conditions defined in this study. Average crystallite size values have been extracted from Rietveld refinement in the maximum XRD peak, for each t_M, see figure 1 (b). It is noted that the crystallite size decreases rapidly from 48 nm for non-milled (NM) sample, to 26 nm for 3 hours milled sample, and then the rate decreases with increasing milling time. An average crystallite size of around 21 nm was obtained for 12 hours of milling, indicating that for this time, is more difficult obtain smaller particle size by ball milling method.

Figure 2 shows the zero field cooling (ZFC) and field cooling (FC) curves of magnetization as a function of temperature for NM sample (a) and for the sample subjected to 12 hours of milling (b), measured during warming up from 5 to 375 K, in 500 Oe after cooling in zero field and applied field of 1 kOe, respectively. M (T) shows the ferromagnetic-paramagnetic (FM-PM) transition typical for this concentration. We observed a difference between ZFC and FC curves, for both NM as for 12 h milled. The curves coincide at temperatures above 250 K and become zero around the Curie critical temperature T_C, where the FM-PM transition occurs. It is worth noting that T_C is similar for both samples approximately at 370 K, and similar to that reported for LSMO-40 in bulk. However, the magnetization at low temperature for non-milled sample, shows that FC at 5 K is around of 0.10 (emu) higher than ZFC, as it is observed in figure 2 (a), and becomes about two times larger than ZFC for the sample with 12 hours of milling, figure 2 (b). This indicates that FC and ZFC curves show a magnetic irreversibility, which manifests below to approximately 250 K, and increases with the milling time.

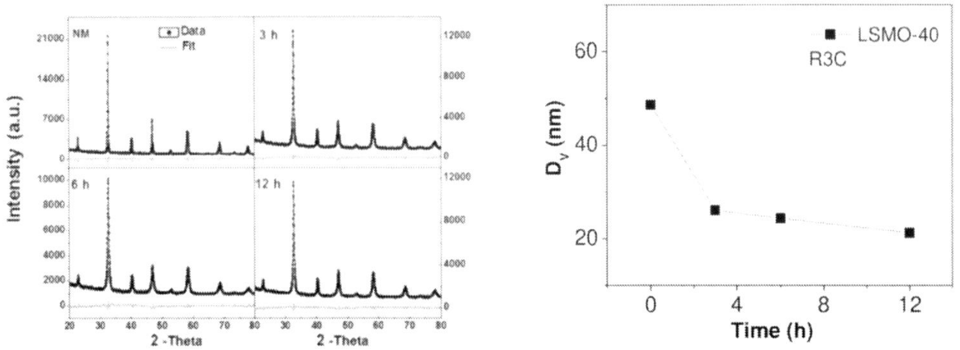

(a) (b)

Figure 1. (a) Experimental and calculated XRD patterns with Rietveld refinement for LSMO-40 with different milling time (NM is the sample without milling). (b) Average crystallite size (D_v) as a function of t_M.

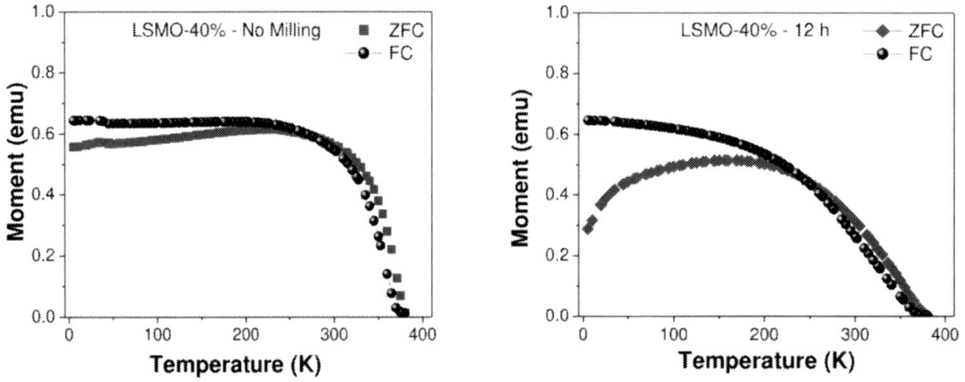

(a) (b)

Figure 2. ZFC and FC magnetization of LSMO as a function of temperature for no milling, NM (a) and for milling time of 12 hours (b).

This behavior can also be attributed to the presence of surface inhomogeneities related with a spin disorder or magnetic anisotropy, which increases when the grain size decreases. Above 250 K the thermal fluctuation balances the anisotropy energy, and in this way both curves coincide around of transition. However, it is observed a broadening of the magnetic transition when increase the t_M, compared with a more abrupt transition in the NM sample, where the behavior is comparable with that of bulk LSMO and the size effects seems that are not present. As to the magnetization as a function of applied magnetic fields M(H), studied in a range of ± 5 kOe, at 5 and 300 K for NM and 12 h samples, are shown in figure 3 (a). From the data one can find that the saturation magnetization varies slightly with the milling time, taken at the same temperature value, but the difference is most evident at low temperature compared with the values at room temperature, around of 0.75 emu for both samples, shown in the figure 3 (a). Nevertheless, the thermal dependence of coercivity decreases with increase of temperature for small grain. The coercive field H_C, measured at 5 K is larger than that value corresponding to 300 K for the sample with 12 h of milling. Figure 3 (b) shows the H_C values for all milling time, taken to 5 and 300 K.

In general, there is an increase in the H_C values with t_M at 5 K, and it became one order of magnitude higher (for 12 h sample) than the value measured in the NM sample. However, for the H_C values at room temperature, the coercivity remain almost constant and there is no a significant dependence with t_M. Such behavior is related with disordered states at the surface level, which influence the magnetic properties of the particles when their size is diminished. In this way, a reduction in the magnetization and increase of coercivity, can be related with a decrease in the spin order in the surface due to size effects that results in a suppression of the DE mechanism, characteristic of these systems.

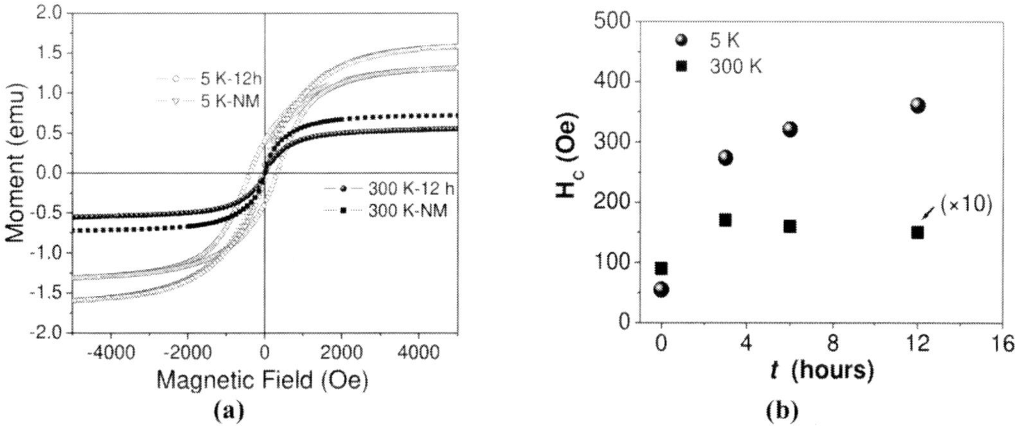

Figure 3. (a) Hysteresis loops for NM and 12 h samples, measured at 5 and 300 K. (b) Coercive fields extracted from M(H) for all milling time, at 5 and 300 K.

4. Conclusions

In summary, the results presented in this work, showed the relation of structural and magnetic properties with decreasing of grain size, of powders of manganite $La_{0.6}Sr_{0.4}MnO_3$ (LSMO-40), subjected to different milling time. The Rietveld analysis showed a single phase present in NM and all samples milled. A decreased average crystallite size values with t_M was found. Samples studied in this work showed dependence with t_M, which was evidenced by decreasing magnetization and increased coercive fields (measured at low temperature). This behaviour is attributed predominantly to size effects which results in a magnetic anisotropy at surface level that can affect the DE mechanism.

Acknowledgements

This work was supported by acta SIU-206 under Sostenibilidad- 2010 strategy, Universidad de Antioquia.

References

[1] Zener C 1951 *Phys. Rev. B.* **82** 403
[2] Salamon M B, Jaime M 2001 *Rev. Mod. Phys.* **73** 583
[3] Bibes M, Barthélémy A 2007 *IEEE Trans. Educ.* **54** 1000
[4] Tokura Y 1999 *Colossal Magnetoresistive Oxides* (Tokyo: Edited by Gordon and Breach)
[5] Hueso L E, Pruneda J M, Ferrari V, Burnell G, Valdés-Herrera J P, Simons B D, Littlewood P B, Artacho E, Fert A, Mathur N D 2007 *Nature* **445** 410
[6] Bongju Kim, Daeyoung Kwon, Jong Hyun Song, Yasuyuki Hikita, Bog G. Kim, Harold Y Hwang 2010 *Solid State Commun* **51** 598
[7] Kameli P, Salamati H, Aezami A 2008 *J. Alloys and Compounds* **450** 7
[8] Dyakonov V, et al 2010 *J. Magn. Magn. Mater.* **322** 3072
[9] Venkataiah G, Prasad V, Venugopal Reddy P 2007 *J. Alloys and Compounds* **429** 1

Anodic stripping voltammetry - ASV for determination of heavy metals

J Barón-Jaimez[1], **M R Joya**[2] **and J Barba-Ortega**[2]
[1] Grupo Plasnamat, Instituto de Investigaciones en Materiales, Universidad Autónoma de México, DF, México.
[2] Departamento de Física, Universidad Nacional de Colombia, Bogotá, Colombia.
E-mail: jairob26@hotmail.com

Abstract. Although voltammetric methods presented a number of difficulties in its early stages, nowadays "ASV" anodic stripping voltammetry is considered one of the most sensitive electro-analytical and suitable for trace-level determination of many metals and compounds in environmental samples, clinical and industrial [1, 2, 3]. Its sensitivity is attributed to the combination of a step of pre-concentration effective together with an electrochemical advanced measurement of accumulated analyte [4]. This paper presents an overview of the voltammetry, which includes a group of electro-analytical methods, in them the information about analyte is obtained from measurements of the current flowing in an electrochemical cell when applied a potential difference to an suitable electrode system.

1. Introduction

Electrochemical methods are based on the measurement of current signals associated with molecular properties or interfacial processes of the chemical species. This due to the direct transformation of desired chemical information (concentration, activity) into a current signal (potential, current, resistance and capacity) according to the method selected [5]. Although there are methods with adequate sensitivity for the determination of heavy metals such as atomic absorption, atomic emission spectrometry and ICP-MS, electrochemical methods are one of the most favorable techniques for determination of heavy metal ions because of their low cost and high sensitivity [6, 7, 8]. Electroanalytical techniques can be divided into two groups, electrodics and ionics; the first "electrodics" taking place at the interface metal-solution and the other "ionics" in the solution bulk (see figure 1(I)) [5]. The majority of electroanalytical methods belong to the field of electrodics.

2. Antecedent

Historically, voltammetry was developed from polarography, which was discovered by chemist Jaroslav Heyrovsky Czechoslovak early 1920. Currently, polarography is only a branch of voltammetry; the earliest methods of voltammetry experienced a number of difficulties [2, 4, 6]. Today, stripping voltammetry is considered the most sensitive electroanalytical technique and suitable for the determination at trace levels of many metals and compounds in environmental samples, clinical and industrial [1, 3]. Its sensitivity is attributed to the combination of a step of pre-concentration effective together with an electrochemical advanced measurement of accumulated analyte [4].

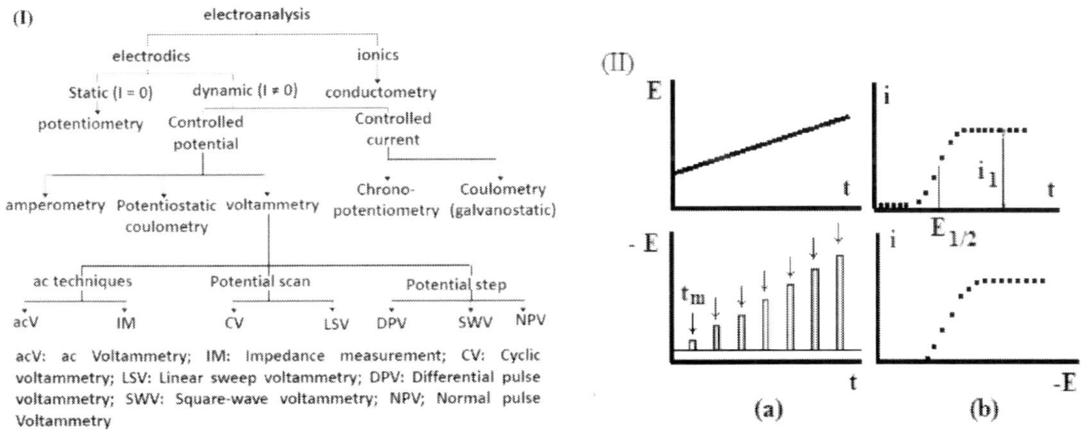

Figure 1. (I) Electro-analysis Methods and (II) two different voltammetry techniques, Linear Sweep Voltammetry (LSV) and Normal Pulse Voltammetry (NPV). a) Principal excitation and b) response curves [5].

3. Voltammetry

All voltammetry can be considered active techniques, because there is an applied potential that promotes a change in the concentration of electro-active species on the electrode surface by reduction or oxidation of the same, they can be described as a function of potential, current and time [2, 9] (see figure 1(IIa)). There are different voltammetry techniques, the difference of these, can provide a variety of chemical information, electrochemistry and physics, such as high quantitative analysis, rate constants for chemical reaction, electrons involved in redox reactions and diffusion constants [2]. We must remember, in the voltammetry there is a current signal produced by two different currents; firstly, corresponding to flow resulting of the analyte oxidation or reduction, this current is called faradic, its magnitude depends of the analytes concentrations in solution and all kinetic steps occurring on the electrode (electron-transfer process), it is based on Faraday's law [5, 9]. The other current is generated by the interface electrode-solution "electric layer double", that produce the capacitive current, it is not related to the electron transfer process, contributing in most cases only to background signals in the measurement analytical. The capacitive current must be separated from the faradic current or at least must be very small for trace analysis, this latter can be accomplished with the use of micro-electrodes and/or may be separated using techniques of step potential (eg. pulse techniques) and the advantages its different characteristics of time of fall [5].

Although there are various voltammetry techniques, the pulse voltammetry is carried out with steps and sweep combinations that improve the speed and detection limits, it can be considered as a family of techniques with adjusted potential, the current is monitored after that the potential has remained constant for a sufficient time, typically 40 milliseconds. The discrimination of the charging current is thus carried out by monitoring the current just before the end of the pulse (or the final step in the case of the stairway), where the load current "capacitive current" is negligible compared with the faradic component [10], (see figure 2(IIa)). The differential pulse voltammetry (DPV) and square wave voltammetry (SWV) improve sensitivity by subtracting the current values monitored at two different times. SWV combines the advantages of the other systems and provides significant improvements; the result is a response similar to a Gaussian signal, characterized by an excellent suppression of the background current and improves the quality of the quantitative information [10].

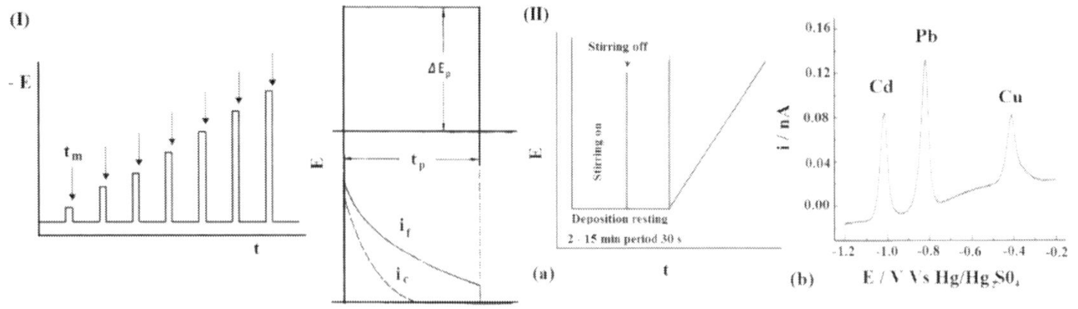

Figure 2. (I) Variation of currents if and ic on the application of pulsed potential [10] and (II) Step of Stripping voltammetry (a) excitation signal (b) response curve [5].

4. Stripping voltammetry

These kinds of techniques are used in the detection and quantitative determination of metals or metal complexes, especially heavy metals in water. The techniques usually consist of a pre-concentration of the metals in the electrode surface, followed by a potential sweep (voltammetry methods) to dissolve pre-concentrated species of interest, making the quantification of them. Depending on the nature of the analyte, different modes of concentration analysis are used; the most common is the Anodic Stripping voltammetry (ASV). The stripping techniques have lower detection limit than any of the commonly used electrochemical techniques, and the preparation of the sample is low, in addition the sensitivity and the selectivity are excellent [2, 6, 9, 10]. His procedure consists of three main steps, (see figure 2(IIb)): Step of pre-concentration: This step provides different modes for pre-concentration of the analyte on the working electrode, which can be summarized by two way, electrochemically (with or without potential stimulus) and by adsorptive. Thanks to this step is achieved the species concentration in volume very much smaller compared to the volume of the solution. To achieve reproducible results is necessary control the hydrodynamic parameters (pre-concentration time, stirring, temperature, electrode area and initial potential applied [4]. The pre-concentration allows a increase of sensitivity of other 2 or 3 orders of magnitude, making it feasible to operate with analyte concentrations 10^{-10} M or even lower, sensitivities comparable with others techniques not electro-analytical characterized by high sensitivity [4]. Step of Resting: After a time perfectly measured, stops electrolysis and stirring, but remains constant initial potential. During this time, is achieved the homogenizing of the substance on the electrode and recovering the diffusion regime [4]. Step of Stripping: During this step, the deposited analyte is determined by a procedure voltammetric by stripping itself, through a potential sweep in the opposite direction to the initial. This sweep can be performed by various techniques (eg, differential pulse, square wave, linear sweep or staircase). The two most commonly used are the differential pulse and square wave, due to their advantages.

These voltammetry techniques can be classified according to the nature of the accumulation process and dissolution in anodic and cathodic tripping voltammetry, stripping potenciometric analysis or adsorptive stripping voltammetry [2, 4, 10]. Anodic Stripping Voltammetry ASV:The voltammetry is now widely recognized as a powerful tool for the analysis of metal ions and certain organic compounds in solutions. The accumulation step covers the reducing of metal cations at a constant potential for several minutes, especially assisted by convection, forming a composite or alloy (bismuth electrodes and mercury). The resulting concentration of the element to be

Journal of Physics: Conference Series **466** (2013) 012023 doi:10.1088/1742-6596/466/1/012023

detected in the electrode is substantially greater than in the solution analyzed due to the volume of the electrode is much smaller in comparison with the volume of solution [5, 10]. After to finish stirring, the potential is changed to a more positive potential by means of a linear sweep voltammetry, differential pulse or square wave like described above, which results in oxidation of the metal the amalgam or alloy, returning again to the solution and recording a current peak due to this process. The current peak height reflecting the detected concentration of the material in the electrode, which is proportional to the amount of material dissolved in the solution as long as they remain appropriate experimental parameters such as the area of the electrode, potential pre-concentration time, time and potential of deposition, conditions of stirring, temperature, etc. The accumulation time necessary depends of the analyte concentration and can reach 20 minutes to a level of 10^{-9} mol dm^{-3}. With this technique ASV, are detected around 15 amalgams or metals alloy depending of the electrode including Ti, Cd, Zn, Cu, Bi, In and Ga. The analysis of traces of other metal ions (Hg, Au, As, Se) may be performed after they are deposited as a layer of electroplated metal in solid electrodes made of carbon or gold [5, 10].

5. Conclusions

Electrochemical methods such as ASV anodic stripping of the hand of a suitable electrode allow us to detect concentrations of heavy metals in the ppm range (single digits), this levels range allow us to be at the maximum limits for human consumption according with the World Health organization, which leads to the conclusion that it is a promise by its low cost, high sensitivity and diversity of this family electrochemical techniques.

Acknowledgements

J. Barón-Jaimez would like to thank Ana Felisa Jaimes by emotional support, J. Barba-Ortega thank to Alejandro Barba and Ely Dannier V-Niño of LNDC-UFRJ by very useful discussions.

References

[1] C Kokinos and I Raptis 2009 *Procedia Chemistry B* **1** 1039
[2] F Settle 1997 *Handbook of Instrumental Techniques for Analytical Chemistry* (Arlington: National Science Foundation)
[3] C Kokinos C and A Economous 2008 *Electrochimica Acta B* **53** 5294
[4] M Amber 1999 *Desarrollo de Tecnicas Electroanalíticas Aplicables a Metales en Fluidos Biológicos de Interés Biosanitario: Determinacin de Cu en LCR, Líquido Cefalorraquídeo* (España: Universidad de Cadiz)
[5] R Cornelis 2003 *Techniques and Methodology B* **5** 429
[6] O Hernandez 2006 *Furoiltioureas: Naturaleza de sus Complejos con $CdCl_2$ y $HgCl_2$ y su Utilización en Sensores Electroquímicos* (España: Universidad de Cadiz)
[7] F Arduini, J Quintana, A Amine, G Palleschi and D Moscone 2010 *Trends in Analytical Chemistry B* **29** 1295
[8] C Linyuan, J Jianbo and W Zhenhui 2008 *Electrochimica Acta B* **53** 2177
[9] H Gunzler, A Williams 2001 *Handbook of Analytical Techniques* (Weinheim: Wiley-VCH Verlag)
[10] C Zosky 2007 *Handbook of Electrochemistry* (Amsterdam: Elsevier)

Structural characterization of mineral with rutile inclusions (TiO_2) and manganocolumbite ($MnNb_2O_6$) by means of X-ray

J F Arcila[1] **and A F Hincapié**[1]
[1] Departamento de Física, Universidad Nacional de Colombia, Bogotá, Colombia.

E-mail: jfarcilaf@unal.edu.co

Abstract. A sample of mineral obtained in the department of Vichada was characterized by the technique of X-ray diffraction in powder samples, in order to determine the crystallographic phases present. After analyzing diffraction patterns, as a result, the mineral had inclusions of Rutile (86.3%) and Manganocolumbita (13.7%). Next, a simulation of the crystal structure of these minerals from the data entered in the databases was carried on. Later, the Rietveld method was used, from this refinement, the new diffraction pattern and the new network parameters were obtained. For the Rutile it was obtained a tetragonal structure, and for the Manganocolumbita an orthorhombic structure was obtained. The characterization is justified among other aspects due to the important applications of these materials in industry and technology, such as Rutile is used in welding rod coatings, in Industries cardboard paper and ink impression among many other uses, the Manganocolumbita is used in special alloys resistant to high temperatures, it also is directly related to the tantalite in coltan formation.

1. Introduction

W. L Bragg accounted for this by regarding a crystal as made out of parallel planes of ions, spaced a distance **d** apart. The conditions for a sharp peak in the intesity of the scattered radiation were: that the X rays should be speculary reflected by the ions in any one plane and that the reflected rays from successive planes should interfere constructively. The path difference between the two rays is just $2d\sin\theta$ where θ is the angle od incidence. For the rays to interfere constructively, this path difference must be an integral number of wavelengths, leading to the Bragg condition: $\lambda = 2d_{hkl}\sin\theta$.

The integer n is know as the order of the corresponding reflection [1]. The Bragg law is a consequence of the periodicity of the lattice [2].

The interatomic distances in crystals and molecules amount to 0.15 - 0.4 nm, which correspond in the electromagnetic spectrum with the wavelength of X-rays having photon energies between 3 and 8 keV. Accordingly, phenomena like constructive and destructive interference should become observable when crystalline and molecular structures are exposed to X-rays. To produce X-rays used in difraction is need a voltage of about 35 kV [3].

In the powder method we put a crystal in an X-ray beam to observe any diffraction: The crystal has to be placed in the beam so as to fulfill the Bragg condition for diffraction. The Bragg-Brentano geometry is the most commonly used geometry for powder diffractometers. The source and the detector are each moved by an angle θ, while the sample is fixed horizontally. Alternatively the source is fixed and the sample is rotated by θ and the detector by $2\,\theta$. The sample is used in reflection mode and a comparably large amount of sample is needed, typically $0.5cm^3$. Due to the large irradiated sample surface. The standard version uses filtered radiation and a monochromator in the diffracted beam. In this way fluorescence radiation is effectively removed, and in general the background level is very low. The secondary monochromator will not remove the $K_{\alpha 2}$ contribution and the peaks gradually split up in two at higher 2θ-angles. The Bragg-Brentano geometry is very good with medium to highly absorbing samples [3].

The Rietveld Method is a method of refining powder diffraction data to find the crystal structure. The practical reason to perform Rietveld refinement is that solving an unknown crystal structure, calculating the amount disorder or mixing on a Wyckoff site, quantitatively determining the percentages and determining the crystallite sizes in the samples. The method fits the whole pattern at once and refines: atomic positions, disorder or mixing between atomic sites, lattice, profile and background parameters. Some factors that affecting peak intensities are: Structure factors, multiplicity, lorentz factor, polarization factor, temperature factor or atomic displacement, absorption or preferred orientation [4].

2. Experimental Procedure

To make an Analysis of crystallographic periods of our mineral, the measurements of the diffraction patterns were done by means of X-ray diffraction, a diffractometer brand X-Pert PRO MPD Panalytical was used. Working at a voltage of acceleration of 45 kV and a current intensity of 40 mA, with monochromatic line K$\boldsymbol{\alpha}$ of copper (Cu-K wave length 1.540598 Å), With a time of 8 s per step and a step size of 0.020 degrees 2$\boldsymbol{\theta}$ in continuous mode, from 10 degrees to 90 degrees 2$\boldsymbol{\theta}$.

The sample was obtained in an agreement with the Departamento de Geologia de la Universidad Nacional de Colombia, this sample had been found in the Department of Vichada embedded in a stone. When it was given to us it was in a solid status and was amorphous, by means of a hammering procedure (using an agate stone) it was obtained a sample in powder of this material that finally was obtained and given to the specialized technician in order to get the diffraction patterns.

3. Results Analysis

Below we will present the main results derived from X-ray characterization, it is important to note that the motivation for the exploitation of this kind of minerals depends on how evident it is, on its inclusions, and also on its possible applications such as technological and industrial applications. In this case, lies essentially the agreement, so the characterization within its purposes shall clarify the useful components, without leaving aside of the interest scientific. The figure 1 related the diffractogram obtained.

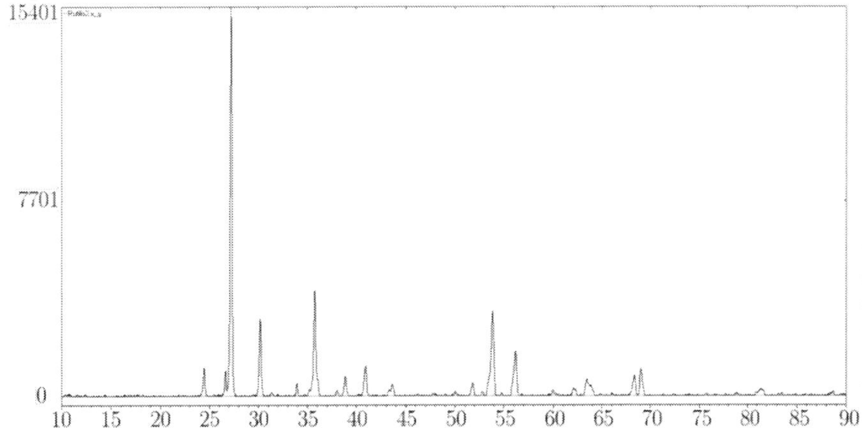

Figure 1. Experimental diffractogram. intensity (AU) Vs 2θ.

Using X-Pert program it could be established that the characterized mineral had Rutile (TiO_2) and Manganocolumbita ($MnNb_2O_6$) inclusions and determines that the sample has 85% Rutile and 15%

Journal of Physics: Conference Series **466** (2013) 012024 doi:10.1088/1742-6596/466/1/012024

Manganocolumbite. These values were obtained by means of a comparison of patterns on each inclusion on program data base.

When minerals were determined by means of mineral database AMCSD it was carried on the crystallographic structure simulation as its diffraction pattern with the PCW program. Next, it is present the Rutile structure and its Manganocolumbite respectively.

A tetragonal structure of Rutile was obtained, the characteristics of this structure are shown below: $\alpha = \beta = \gamma = 90°$ and $a = b \neq c$.

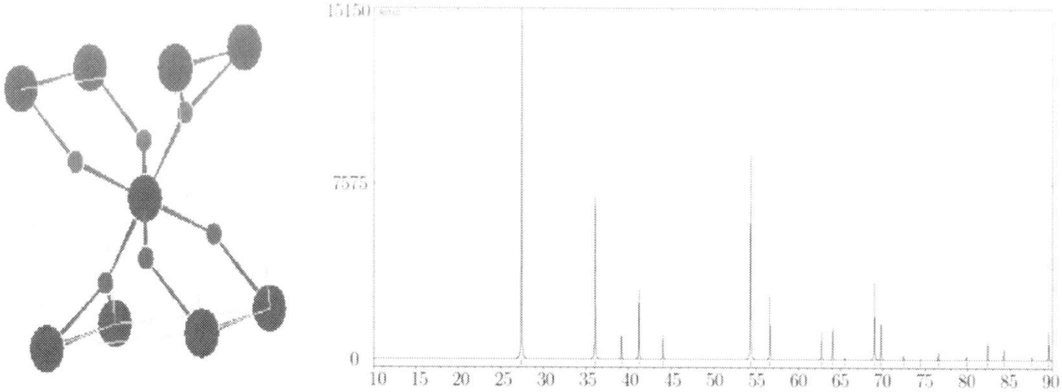

Figure 2. Crystal structure of rutile and the diffractogram simulated.

For Manganocolumbita we got an orthorhombic structure, the relationship of angles and sides, and it is the following one: $\alpha = \beta = \gamma = 90°$ and $a \neq b \neq c$.

The refinement Rietveld was done in order to get our sample was obtained and the definitive diffraction pattern was obtained (see figure 4).

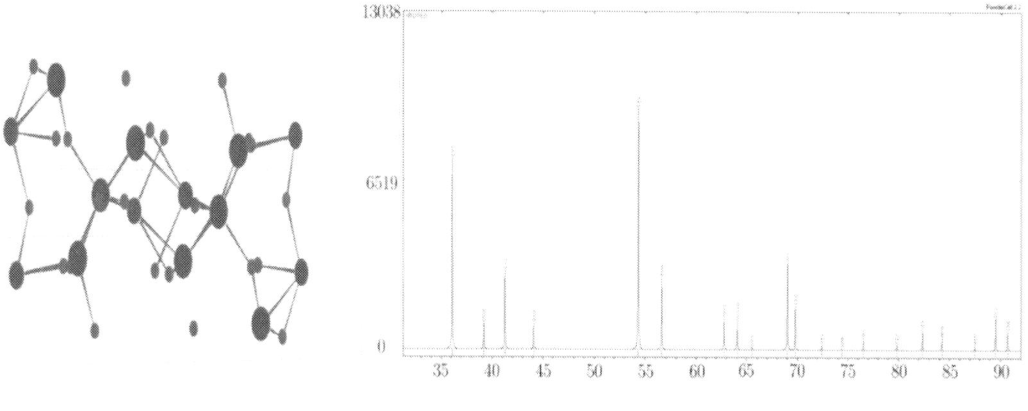

Figure 3. Crystal structure of manganocolumbite and the diffractogram simulated.

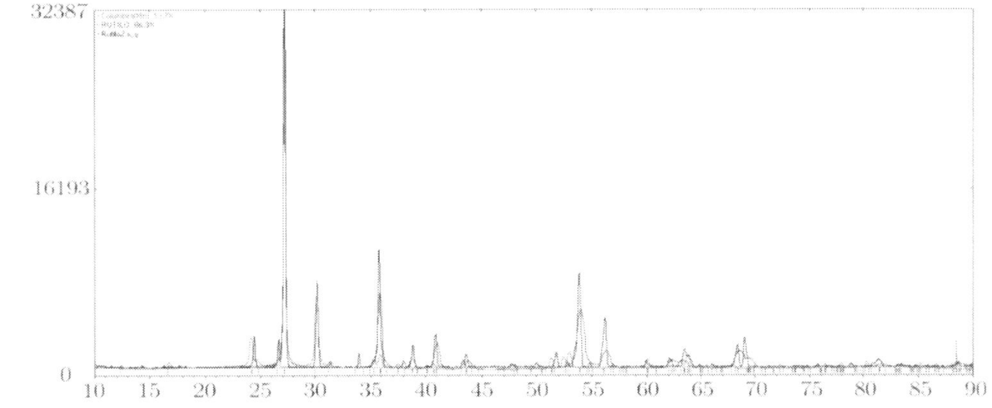

Figure 4. Rietveld refinement.

Finally presents abstract the lattice parameters in the next data table:

Table 1. The lattice parameters.

	Rutile (TiO_2)			Manganocolumbite ($MnNb_2O_6$)		
	Experimental	Database	Rietveld R	Experimental	Database	Rietveld R
a(Å)	4,6255	4,593	4,6373	14,246	14,352	14,306
b(Å)	4,6255	4,593	4,6373	5,726	5,74	5,728
c(Å)	2,959	2,959	2,9854	5,093	5,09	5,08

Since we know the components isolated from the mineral, we can conclude that the characterization among other concepts is justified by the important applications of these materials in industry and technology, for example Rutile is used in welding rod coatings, in Industries cardboard paper and ink, impression among many other uses, the Manganocolumbita is used in special alloys resistant to high temperatures, it also is directly related to the tantalite in coltan formation.

Due to these materials different applications, it will be of great interest, firstly the exploitation of this mineral, and on the other hand, the separation into its components, for the separation of the material we suggest to take advantage of the magnetic properties of each inclusion.

4. Conclusions and remarks

After Rietveld refinement, the following lattice for the Rutile parameters were obtained a= 4.6373 Å, b = 4.6373 Å, c = 2.9854 Å. The Manganocolumbita parameters were a = 14,306 Å, b = 5,728 Å, c = 5.08 Å. For other hand the characterization is justified among other concepts by the important applications of these materials in the industry and technology, also is possible to think about the magnetic separation as one separation option in components of our material.

References

[1] Ashcroft N W and Mermin N D 1976. *Solid State Physics* (Orlando: Saunders)
[2] Kittel C 1996 *Introduction to Solid State Physics* (New York: Jhon Wiley & Sons, Inc.)
[3] Birkholz M 2006 *Thin Film Analysis X-ray Scattering* (Germany: Willey-UCH)
[4] Rietveld H 1969 *J. appl. Cryst.* B **2** 65

Journal of Physics: Conference Series **466** (2013) 012025 doi:10.1088/1742-6596/466/1/012025

Bismuth electrodes, an alternative in stripping voltammetry

J Barón-Jaimez[1], **M R Joya**[2] **and J Barba-Ortega**[2]

[1] Grupo Plasnamat, Instituto de Investigaciones en Materiales, Universidad Autónoma de México, DF, México.

[2] Departamento de Física, Universidad Nacional de Colombia, Bogotá, Colombia.

E-mail: jairob26@hotmail.com

Abstract. The heavy metals are known as highly toxic contaminants, the processes carried out in industry contribute that finally they remain dispersed in effluents and sewage, doing part of the food chain. The importance of controlling the levels of these heavy metals has become an international policy, so it has generated interest in developing new analytical methodologies for its determination [1, 2, 3, 4]. The stripping voltammetry has been considered as a family of electro-sensitive analytical techniques useful for the determination of trace levels of many metals in environmental, clinical and industrial samples [3, 4]. This work presents an overview of these bismuth-based electrodes which were introduced around 2000, which have interesting characteristics for detection of heavy metals and which represent an alternative to mercury electrodes.

1. Introduction

In the last years, the industrialization has come accompanied by different processes, which lead that the heavy metals remain finely dispersed in water, air and soil, as part of the our food chain [1]. Although there are methods with adequate sensitivity for the determination of heavy metals such as atomic absorption, atomic emission spectrometry and $ICP-MS$, electrochemical methods are one of the most favorable techniques for determination of heavy metal ions because of their low cost and high sensitivity [5]. In analytical chemistry, Anodic Stripping Voltammetry techniques (ASV) are a promising alternative to spectroscopic analysis techniques, particularly by their high sensitivity "*ppb*" in the detection of the elements trace in fluid and that the sensors are easily coupled to field units in addition of its low cost, low volume of reagents consumed and can be reused for several measures [6]. In this area, the potentiometric sensors represent a very interesting area and with expansion, their great development and acceptance is due to its high sensitivity and the ability to make direct measurements or act as indicators electrodes in valuation, such measures are no affected by the color or turbidity of the sample or analysis solution [7, 8, 9].

2. Results

Although mercury electrodes have been widely used for this type of electrochemical techniques because of its high reproducibility and sensitivity [9], its toxicity makes it less and less popular [3, 9, 10, 11, 12]. Different studies have been assigned to the construction of electrochemical

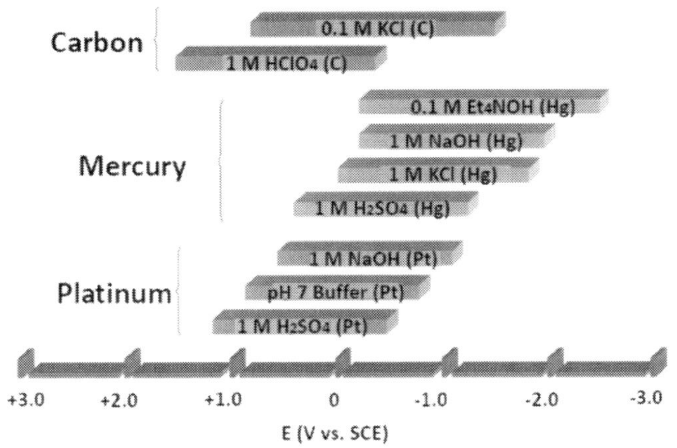

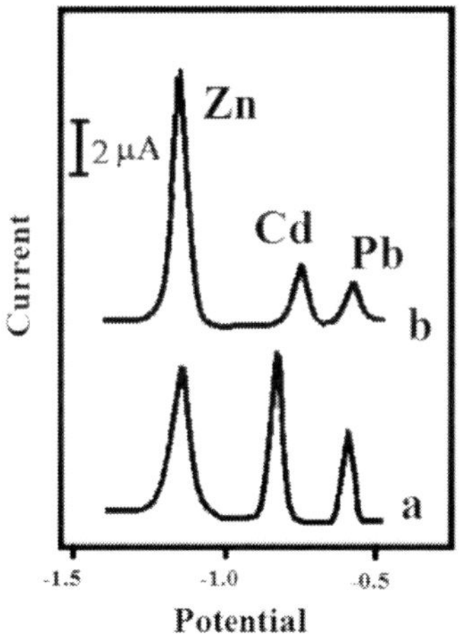

Figure 1. (Left) Behaviour potential windows for different materials. (Right) Current peaks of Cd, Zn and Pb for voltammetry. a) Bismuth electrode b) Mercury electrode (14).

sensors that do not contain mercury, a variety of different materials have been suggested and explored to produce electrodes of work for this purpose, such as noble metals (Pt, Pd, Au, Ag) and other metals $(Ru, Cu, Co, Ni, Pb, Sb, Bi, Al)$ [7, 13]. However, bismuth is chosen due to its low toxicity [3, 9, 11, 14, 15] wide potential window and ability to form alloys with different metals [9, 10, 16, 17]. This physical-chemical behavior allows its use in the simultaneous detection of multiple electro-active species present in the same sample.

3. Bismuth electrodes in stripping voltammetry

The stripping voltammetry has been considered as a family of electroanalytical techniques more sensitive and useful for determining the level of trace metals in many samples of environmental origin, clinical and industrial, as well as organic components. The risk associated with mercury electrodes is their use, handling and disposal of waste because of its toxicity, this situation generated different researches to overcome these drawbacks; so, around 2000 the electrodes BiFEs (Bismuth Film Electrodes) were introduced, they are fabricated of a bismuth layer deposited on a suitable substrate [3, 4], different materials have been used as electrode substrate such as carbon, glassy carbon, carbon fibers, carbon paste, graphite, graphite impregnated with wax, gold, platinum [10, 16]. These Bismuth electrodes are a promise for replacing the mercury electrodes in stripping voltammetry, as well as their low toxicity, the BiFEs have shown to have electro-analytical properties similar to the mercury electrodes, such as their wide potential window, in addition to its ability to form alloys with various metals, their partial insensitivity to dissolved oxygen, their signals well defined and simple preparation [3]. The current peaks obtained at the voltammograms when uses bismuth electrodes tend to be sharp and well defined [14], in figure (1, right), which enables the identification and quantification of the metals present in the water to reliably way, quickly and economic, could be performed even in the same site where we take the sample. In recent years, different types of modified bismuth and bismuth electrodes have been developed showing a behavior comparable to MFEs (Mercury Film Electrodes) in ASV, reported methods for preparing electrodes with an active surface of bismuth have been fabricated by modifying of volume the surface of the solid electrode with metallic bismuth such as mixtures of powders of bismuth, bismuth oxide and carbon [14, 18], thermal evaporation [19], sputtering [4, 11] and electro-deposits compound Bi (III) both ex situ [10, 12, 20, 21] as in situ [9, 10, 17, 15, 21, 22, 23, 24, 25] in a conductive material. The

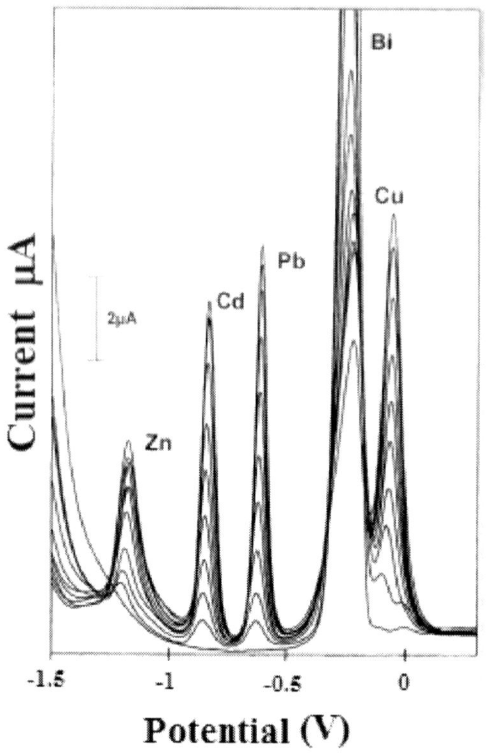

Figure 2. Square-wave stripping voltammograms for increasing concentration of copper, lead, cadmium and copper (21).

method most widely used for the preparation of BiFEs is electrochemical performed by static potentiometry (in situ or ex situ) and occasionally, by galvanostatic method, cyclic potential or pulse amperometry. However deposits by electrochemical means involving the use of salts of Bi (III), which can complicate experimental procedures and requires a conductive substrate [3, 4]. There are several reports in the literature of bismuth based electrodes were tested for the determination of trace metal ions and occasionally for the detection of organic compounds using voltammetric techniques in various forms. Some of the metals that have been reported so far include $Cd, Pb, Zn, Tl, In, CuNi, Co, Fe, Al, Mo, V, Ti, U$ and Cr [10, 26]. Furthermore, metals have been detected in trace amounts in a variety of biological samples (hair, urine and blood) [10, 26] and environmental samples (air, tap water, sea water, river water, drains and water soils) [7, 14, 16], foodstuffs (tea, cabbage, celery, spinach, wine and tomato sauce) [7, 10, 14, 15] and Petrol [7].

4. Conclusions

In the last decade, bismuth electrodes manufactured by different methods and due to their physico-chemical Characteristics are a promise for stripping voltammetry in the determination of heavy metals especially, these electrodes are bringing us more and more to the determination of a single digit ppb of these elements, which in most cases are the limits permitted by international health organization for human consumption.

Acknowledgements
The authors would like to thank Sandra E. Rodil of the Plasnamat and Ely Dannier V-Niño of LNDC-UFRJ by very useful discussions.

References

[1] www.auroville.org. *Auroville innovative Urban Management IND-015* 1 annexes
[2] Foro Interguvernamental de Seguridad Quimica - IFCS 2006 Budapest
[3] C Kokkinos, I Raptis, A Economou and T Speliotis 2009 *Procedia Chemistry* **1** 1039
[4] C Kokinos and A Economous 2008 *Electrochimica Acta* **53** 294
[5] O Hernandez 2006 *Furoiltioureas: Naturaleza de sus Complejos con $CdCl_2$ y $HgCl_2$ y su Utilización en Sensores Electroquímicos* (España: Universidad de Cadiz)
[6] I Sierra 2009 http://www.madrimasd.org/informacionidi/noticias/noticia.aspid=38007
[7] F Arduini and J Quintana 2010 *Trends in Analytical Chemistry* **2-11** 1295
[8] M J Gismera 2006 *Desarrollo de Nuevos Sensores Potenciometricos para Metales Basados en Receptores con Grupos Tio y Ditio* (Madrid: Universidad Autónoma de Madrid)
[9] L Cao and J Jia 2008 *Electrochimica Acta* **53** 2177
[10] A Economou 2005 *Trends in Analytical Chemistry* **24-4** 334
[11] C Kokkinos, A Economou 2007 *Electrochemistry Communications* **9** 2795
[12] E Hutton, B Ogoreve, S Hocevar, F Weldon, M R Smyth and J Wang 2001 *Electrochemistry Communications* **3** 707
[13] R Pauliukaite, S Hocevar, B Ogorevc and J Wang 2004 *Electroanalysis* **16** 719
[14] J Wang 2005 *Electroanalysis* **17** 1341
[15] H Xu, L Zeng 2008 *Food Chemistry* **109** 834
[16] S Hocevar S, I Svancara, K Vytras and I Svancara 2005 *Electrochemica Acta* **51** 706
[17] L Baldrianova, I Svancara, M Vicek, A Economou and S Sotiropoulos 2006 *Electrochimica Acta* **52** 481
[18] L Kumari, S Lin, J Lin, Y Ma, P Lee and Y Liou 2007 *Applied Surface Science* **253** 5931
[19] N Serrano, A Alberich, J M Diaz-Cruz, C Arino and M Esteban 2008 *Electrochemica Acta* **53** 6616
[20] I Svancara, L Baldrianova, M Vicek, R Metelka and K Vytras 2005 *Electroanalysis* **17** 120
[21] J Wang, J Lu, U Kirgoz, S B Hocevar and B Ogorevc 2001 *Analytical Chimica Acta* **434** 29
[22] C Prior, C E Lenehan and G S Walker 2007 *Analytical Chimica Acta* **598** 65
[23] C H Xiong, H Q Luo and N B Li 2011 *Journal of Electroanalytical Chemistry* **651** 19
[24] Y Wu, N B Li and H Q Luo 2008 *Sensor and Actuator B: Chemical* **133** 677
[25] Z Zou, A Jang, P Wu, J Do, J Han, P Bishop and C Ahn 2007 *Eleventh International Conference on Miniaturized Systems for Chemistry and Life Sciences* **251**
[26] R Cornelis 2003 *Handbook of Elemental Speciation: Techniques and Methodology* (England: Wiley)

Journal of Physics: Conference Series **466** (2013) 012026 doi:10.1088/1742-6596/466/1/012026

Exciton complexes in impurity doped quantum ring

L Porras[1] and I Mikhailov[1]
[1] Escuela de Física, Universidad Industrial de Santander, Bucaramanga, Colombia.

E-mail: luis.c.porras.m@gmail.com

Abstract. We propose a simple unified method for theoretical description of free and bound to impurity neutral and charged excitons, confined in a heterostructure with ring-like geometry. In order to assess the experimentally relevant domain of parameters, we adopt a model of a narrow ring, when for both neutral and charged excitons 3D wave functions can be separated. By using the Fourier series method, we have calculated the energy spectra of excitons complexes in a quantum ring as a function of the electron-to-hole mass ratio, and the magnetic field strength. The size-dependent magnetic oscillations of energy levels of excitons' complexes spectra have been revealed.

1. Introduction

Very recently, a number of experimental and theoretical studies of neutral X, negatively X^- and positively X^+ charged excitons confined in quantum dots have been performed [1]. The possibility to control individual excitations, in quantum dot structures is an indispensable prerequisite for the new field of quantum information processing in a semiconductor environment, in which exciton complexes are considered as the intermediate states. Fabrication of nanometer-sized lithography defined semiconductor quantum rings (QR) generated interest in the Aharonov-Bohm (AB) effect induced by charged excitons [2]. As it was shown in [3], the AB effect is more pronounced under condition that the electron and hole move in a ring-like structure over different paths providing a non-zero angular momentum. It is clear that the separation between their paths is strongly related to the electron-to-hole mass ratio, the larger the ratio, the more distant are paths. There is an additional reason for studying the effect of the masses ratio, associated with a possibility of the participation in the formation of the exciton complexes of particles with different masses, such as light-hole and heavy-hole, or as a case of extremely heavy carriers a donor or an acceptor. Therefore, below we propose a simple method for solving the Schrödinger equation for a particular case when exciton complexes consist of one electron and two holes with arbitrary masses m_{h1} and m_{h2} confined in a narrow QR that allows us to analyze in the framework of the same formalism the properties of different types of the excitonic complexes such as free trions X^+ with participation of holes with different masses, free exciton when $m_{h2} \to 0$, or exciton bound to acceptor when $m_{h2} \to \infty$. We apply the formalism for analyzing the AB effect for free and bound to an acceptor X and X^+ complexes confined in QR and show that a decrease of the electron-to-hole effective mass ratio provides a quenching of the AB oscillations of the lower energy levels when a free exciton is bound to an acceptor, similarly to one observed recently for QRs with single and coupled donors [4].

2. Theory

In order to assess the experimentally relevant domain of parameters, we adopt a simple model of a narrow ring, in which its height h and width w are significantly smaller than its centerline radius R. In this case, which we call the adiabatic limit, 3D wave equations for exciton complexes can be separated and the analysis of their low-lying states is reduced to a simpler problem of 1D rotation of two holes with effective masses $m_{h1}*, m_{h2}*$ and the electron with mass m_e*, whose positions are given by three polar coordinates J_{h1}, J_{h2} and of J_e, respectively [5]. By using the exciton effective Bohr radius $a_0^* = \text{h}^2 e / me^2$ as the unit of length, the effective Rydberg $Ry^* = e^2/2ea_0^* = \text{h}^2/2ma_0^{*2}$, as the energy unit and $g = e\text{h}B/2mcRy^*$, as the unit of the magnetic field strength, being $m = m_e^* m_{h1}^* / (m_e^* + m_{h1}^*)$ the reduced mass, the Hamiltonian which describes simultaneously a free exciton for $\eta = m_{h2}*/m_{h1}* = 0$, free trion X^+ for $0<\eta\leq 1$ and exciton bound to an aceptor for $\eta = m_{h2}*/m_{h1}* \to \infty$ can be written as:

$$H(\eta) \approx (2+\eta)\left[\frac{\pi^2}{h^2}+\frac{\pi^2}{w^2}\right]-\frac{1}{I_h}\left[\frac{\partial^2}{\partial\vartheta_{h1}^2}+\eta\frac{\partial^2}{\partial\vartheta_{h2}^2}\right]-\frac{1}{I_e}\frac{\partial^2}{\partial\vartheta_e^2}+\frac{\gamma^2\bar{R}^2}{4}\left[\frac{1+\eta}{\mu_h}+\frac{1}{\mu_e}\right]$$

$$+i\frac{\gamma}{\mu_h}\left[\frac{\partial}{\partial\vartheta_{h1}}+\eta\frac{\partial}{\partial\vartheta_{h2}}\right]-i\frac{\gamma}{\mu_e}\frac{\partial}{\partial\vartheta_e}+\lambda V_c(\vartheta_{h1}-\vartheta_{h2})-V_c(\vartheta_{h1}-\vartheta_e)-\lambda V_c(\vartheta_{h2}-\vartheta_e). \quad (1)$$

Here the parameter λ is equal to zero for excitons and to one for trions. Additionally, in the expression (1), obtained in framework of the adiabatic approximation [6], are used the following notations $\bar{R}^2$ » $R^2 + w^2/12; V_c(J)$ » $2/\sqrt{w^2/4+\bar{R}^2\sin^2 J/2}$, and $m_i = m_i^*/m;\ I_i$ » $m_i\bar{R}^2;\ \ i = e,h$.

In the center-of-mass coordinates

$$Q = (I_h(J_{h1}+hJ_{h2})+I_eJ_e)/I;\ J_1 = J_{h1}-J_e;\ J_2 = J_{h2}-J_e;\ I = (1+h)I_h + I_e \quad (2)$$

The Hamiltonian (1) is separated completely in the case of the exciton ($\eta = 0$) and partially for trion ($\eta = 1$). Being $M = 0, \pm 1, \pm 2, \text{K}$ the angular momentum corresponding to the center-of-mass rotation, the eigenfunctions of the Hamiltonian (1) can search in the form:

$$H(h)\text{Y}(Q,J_1,J_2,h) = E(h)\text{Y}(Q,J_1,J_2,h);\ \text{Y}(Q,J_1,J_2,h) = e^{iMQ}\text{F}(J_1,hJ_2) \quad (3)$$

Here unknown envelope function F describes only the relative particles motion. By using definitions (1) and (2) in equation (3) one can derive after simple algebraic manipulations the following differential equation for the envelope function F :

$$-\frac{1}{I_\mu}\left[\frac{\partial^2\Phi}{\partial\vartheta_1^2}+\eta\frac{\partial^2\Phi}{\partial\vartheta_2^2}\right]+i\gamma\left[\frac{\partial\Phi}{\partial\vartheta_1}+\eta\frac{\partial\Phi}{\partial\vartheta_2}\right]-\frac{2\eta}{I_e}\frac{\partial^2\Phi}{\partial\vartheta_1\partial\vartheta_2}+V_{Coul}(\vartheta_1,\vartheta_2)\Phi = (E-E_0)\Phi;$$

$$E_0 = (2+\eta)\left[\frac{\pi^2}{h^2}+\frac{\pi^2}{w^2}\right]+\frac{M^2}{I}+\frac{\gamma^2\bar{R}^2}{4}\left[\frac{1+\eta}{\mu_h}+\frac{1}{\mu_e}\right]-\gamma M\frac{(1+\eta)I_hm_e^*-I_em_h^*}{[(1+\eta)I_h+I_e](m_h^*+m_e^*)} \quad (4)$$

Here we denote by $V_{Coul}(J_1,J_2) = lV_c(J_{h1}-J_{h2})-V_c(J_{h1}-J_e)-lV_c(J_{h2}-J_e)$ the total Coulomb interaction energy and $I_m = I_hI_e/(I_h+I_e)$ the reduced electron-hole moment of inertia. In our numerical work we solve the equation (4) by using the single and the double Fourier series expansion methods for the exciton and the trion, respectively.

3. Results

We analyse the influence of the magnetic field on the energies of exciton complexes in a ring with material parameters typical for InAs material, $\varepsilon = 12{,}71$, $m_e^* = 0.026m_0$, $m_h^* \approx 0.026m_0$ for light-hole and $m_h^* \approx 0.4m_0$ for heavy hole. Below we present in figures 1 and 2 some results for low-lying energies of neutral exciton ($h = 0, l = 0$), formed by one electron and one light-hole, positively charged exciton complexes formed by one electron and two light ($h = 1, l = 1$) or two heavy holes ($h \approx 16, l = 1$) and also for light-hole exciton bound to an acceptor ($h \to \infty, l = 1$). In all cases we consider a ring with the centerline radius $R = 5a_0^*$, height $h = 0.2a_0^*$ and width $w = 0.2a_0^*$. In figure1 we show some low-lying levels as functions of the magnetic field for charged excitons, formed by one electron and a pair of heavy or light-holes.

It is seen that in both cases the energy levels are distributed in three groups with different responses to the variation of the external magnetic field. The lower group is formed by doubly degenerated levels whose energies are remarkably spaced and almost insensitive to any variation of the magnetic field. The states of this group remain bound in the presence of the magnetic field however the energy levels corresponding to these states suffer a weak double splitting due to exchange symmetry presented in the three-particle complexes. On the contrary, in the intermediate part of the energy spectrum one can observe typical for AB effect oscillations and reordering of energy levels with multiple crossovers between them. The appearance of the AB oscillations for heavy- and light-hole trions is different. In the first case, when the electron mass is 16 times less than the mass of holes, the crossovers are multiple while in the second case, when the masses of all three particles are similar, there is additional symmetry and the multiple crossovers split in doublet's.

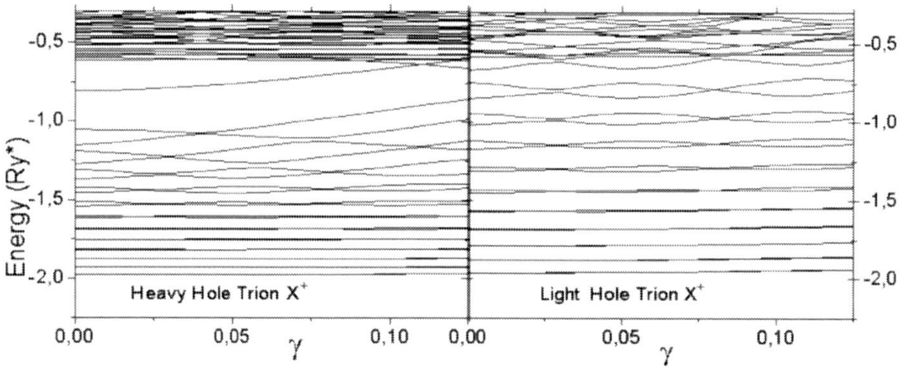

Figure 1. Energies of lower levels of charged excitons, formed by one electron and a pair of heavy or light holes confined in quantum ring, as functions of the magnetic field.

Both a weak double splitting and the multiple crossovers of the lower energy levels in figure1 we attribute to interaction of the tunnel currents, generated by relative motion of the three-particle system through inter-particle interaction potential, with the external magnetic field. The higher the system energy the larger is the probability for the tunnelling and the upper is the amplitude of the AB oscillation. When the energy increasing becomes higher than barrier of the interaction potential the particles relative motion is transformed in a free rotation. Corresponding to these states energies dependencies in the upper part of figure 1 describe an independent rotation of three particles in their unbound state. Therefore the upper energies form quasi-continuum levels with multiple crossovers and reordering between them.

In figure 2 we compare energies of (a) the unbound light-hole exciton and (b) the exciton bound to an acceptor in quantum ring, as functions of the magnetic field. It is seen a vast difference between the energy spectrum of a free exciton and an exciton bound to an acceptor. The exciton bound to an acceptor loses almost completely its individual properties. One can observe in figure 2 (a) that the unbound exciton has only few energy levels insensitive to the external magnetic field with usual for

Coulomb attraction a large spacing between the ground and the first excited states while the upper part of the exciton energy spectrum shows a typical for AB oscillation dependence on the external magnetic field. On the contrary, the exciton bound to an acceptor have many low lying levels that are completely independent on the external magnetic field. Also, comparing corresponding curves in figure1 and 2, one can see that there is a general similarity between energy spectra of the heavy-hole trion and the exciton bound to an acceptor, but the energies in the second case are slightly lower.

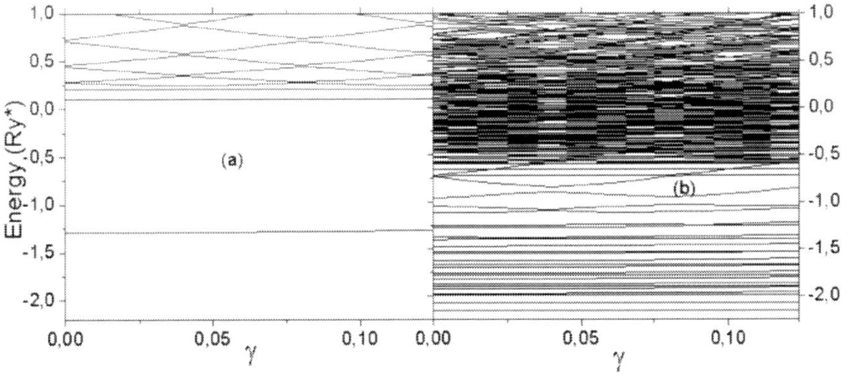

Figure 2. Energies of lower levels of (a) unbound and (b) bound to an acceptor the light-hole exciton confined in quantum ring, as functions of the magnetic field

4. Conclusion

A unified differential equation that describes in the adiabatic limit the energy spectra of two and three carriers confined in narrow quantum ring is derived and the double Fourier series expansion method is applied in order to analyze the Aharonov-Bohm oscillations of the lower energy levels in the presence of magnetic field. The energies of low-lying states of the neutral X, positively heavy- and light–hole charged excitons X^+, and exciton bound to an acceptor, as functions of the magnetic field strength have been calculated. A high sensibility of the energy spectrum on the electron-to-hole mass ratio has been revealed. It is shown that a decrease of this ratio conduces to quenching of the oscillation of the lower energy levels under the variation of the external magnetic field.

Acknowledgements

We wish to acknowledge to High Performance Computing Center from Universidad Industrial de Santander - UIS for its support in the program execution. This Work was financed by the «DIEF de Ciencias, UIS» and the «Programa Nacional de Formación de Investigadores 2011 (Cod: 528)» subscribed with Colciencias, Colombia, by mean of the grant RC - No. 275-2011.

References

[1] Peter Michler 2003 *Single quantum dots: fundamentals, applications and new concepts* (Germany: Springer-Verlag Heidelberg)
[2] M Bayer, M Korkusinski, P Hawrylak, T Gutbrod, M Michel and A Forchel 2003 *Phys. Rev. Lett.* **90** 186801
[3] F Ding, N Akopian, B Li, U Perinetti, A Govorov, F M Peeters, C C Bof Bufon, C Deneke, Y H Chen, A Rastelli, O G Schmidt and V Zwiller 2010 *Phys. Rev.* B **82** 075309
[4] W Gutiérrez, L F García and I D Mikhailov 2010 *Physica* E **43** 559
[5] I D Mikhailov, L F García and J H Marín 2008 *Microelectronics Journal* **39** 378
[6] F M Peeters and V A Schweigert 1996 *Phys. Rev.* B **53** 1468

Journal of Physics: Conference Series **466** (2013) 012027 doi:10.1088/1742-6596/466/1/012027

Dirichlet boundary condition for the Ginzburg-Landau equations

J Barba-Ortega[1], J Barón-Jaimez[2] and M R Joya[1]

[1] Departamento de Física, Universidad Nacional de Colombia, Bogotá, Colombia.
[2] Instituto de Investigaciones en Materiales, Universidad Autónoma de México, DF, México.

E-mail: `jjbarbao@unal.edu.co`

Abstract. As is well known, the Ginzburg Landau phenomenological theory described with a good accuracy the thermodynamic properties of a superconducting material. The system of two coupled nonlinear differential equations is completed with the usual Neumann boundary condition as long as is considered a superconductor insulator interface. In this paper, we solve the Ginzburg Landau equations for a circular geometry containing a half-circular pillar defect and considering the unusual superconducting Dirichlet boundary condition. This choice, leading to take the extrapolation de Gennes length equal to zero. Our results point that, the thermodynamic critical fields, magnetization, free energy and vorticity, depend on the chosen boundary condition.

1. Introduction

Physics properties in a conventional superconducting material has generate much activity in the wide scientific community. Several topologies as samples with surface roughness or surface defects have attracted attention as potential new components for low-temperature electronics physics [1, 2, 3, 4, 5]. Also, many phenomena of current interest are directly related to the non-monotonic interaction between vortices, being a possible explanation for highly debated unusual vortex configurations observed in MgB_2 and $Ba(Fe_{0.95}Co_{0.05})_2As_2$ [6, 7]. The transparent physical ideas behind the Landau theory, as well as the long history of its successful applications, have led to a general belief that captures the physics of the superconducting matter in a much wider range of physically plausible situations. By solving the first non-linear time dependent Ginzburg-Landau equation considering the Dirichlet boundary condition for the superconducting pseudo-function wave, we calculate the thermodynamic critical fields, magnetization, Gibbs free energy and vorticity for a thin disk containing a half-circular pillar defect in presence of a external magnetic field.

2. Theoretical formalism

Let us consider a thin superconducting disk with a half-circular pillar defect, immersed in an insulating medium in the presence of a perpendicular uniform magnetic field $\mathbf{H_0}$. Superconducting matter is described in the Ginzburg-Landau theory by the order parameter ψ, $|\psi|^2$, describes the spatial variation of the superconducting electrons, and the potential vector $\mathbf{A}$, related to the magnetic induction as $\mathbf{h} = \nabla \times \mathbf{A}$, in absence of external currents and using the London gauge ($\nabla \cdot \mathbf{A} = \mathbf{0}$), electrical potential ($\varphi = 0$) (for more details, see [8, 9, 10]) as:

$$\frac{\partial \psi}{\partial t} = -(i\nabla + \mathbf{A})^2\psi + (1-T)\psi(1-|\psi|^2) \quad (1)$$

$$\frac{\partial \mathbf{A}}{\partial t} = (1-T)\mathrm{Re}\left[\bar{\psi}(-i\nabla - \mathbf{A})\psi\right] - \kappa^2\nabla \times \mathbf{h} \quad (2)$$

The dynamical equations are complemented with the appropriate boundary conditions for the order parameter (See [11]):

$$(-i\nabla - \mathbf{A})\psi \cdot \mathbf{n} = \frac{i\hbar}{b}\psi|_n \Rightarrow \begin{cases} b \to \infty \Rightarrow \nabla\psi|_n = 0 & \text{Neumann Boundary Condition} \\ b = 0 \Rightarrow \psi|_n = 0 & \text{Dirichlet Boundary Condition} \end{cases} \quad (3)$$

$\hat{n}$ is the unity vector perpendicular to the surface of the superconductor. b is the deGennes parameter, (see [1]). We will neglect the z-dependence on the order parameter. This is reasonable for thickness of the disk d much smaller than the coherence length. Then $\mathbf{H_0} = \nabla \times \mathbf{A_0}$ inside the disk and the equations 1 can be written as:

$$\frac{\partial \psi}{\partial t} = -(i\nabla + \mathbf{A_0})^2\psi + (1-T)\psi(1-|\psi|^2) \quad (4)$$

$b \to \infty$ leads to the usual Neumann boundary condition for superconducting materials, i.e. $\hat{n} \cdot \nabla\psi = 0$, and $\psi|_n \neq 0$ typical condition for a superconducting/vacuum interface, it will be assumed that normal current density not vanishes at the interface. Now, positives (negatives) values for b simulates a superconductor/metal (superconductor/superconductor) interfaces, for these conditions we have always $\nabla\psi \neq 0$ and $\psi \neq 0$ in the surface of the sample. $b = 0$ simulates a high density of defects in the interface, its lead to take $\psi|_n = 0$ in equation 3, the usual Dirichlet boundary condition in quantum mechanics but unusual for Ginzburg Landau equations for the superconductivity [11, 12].

3. Results and discussion

The parameters used in our numerical simulations for a thin film of pure Al of thickness d (assuming $\xi(0) = 1.2\ \mu m$, $T_c = 1.7K$, $T = 0.85$, the radius of the disk was $R = 8.0$ with a internal radius $r_i = 0.3$, the defect position was $P(r,\theta) = 1.45$ for $\pi \leq \theta \leq 2\pi$ and $r = R/2$ ([13, 14]). Figures 1(c, g) shown the contour plot of the order parameter for several vorticities

Table 1. Thermodynamics values for Dirichlet and Neumann conditions.

B. Condition	$-G(H_{01})$	$-4\pi M(H_{01})$	$N\Phi_0(H_{01})$	H_{01}	H_{02}	H_{03}
Dirichlet	0.815	0.820	2.0	0.650	0.820	1.029
Neumann	0.705	0.712	1.0	1.077	1.05	1.740

(in logarithmical scale (a, e)) and its phase (b, f), also the magnetic induction (d) and (h) the super-current for a disk with a defect for a magnetic field $H_0 = 0.888$ with $N = 6$ vortices (up) and $H_0 = 0.652$, $N = 4$ vortices (down); $N = 2$ vortices remains in the superconductor region in both cases, while four (up), and two (down) sit in the center of the disk. Although they are not visible in the contour plot of the magnitude of order parameter, there is a change in the phase around the hole equal to $\Delta\Theta = 12\pi$ (figure 1(b)), and $\Delta\Theta = 8\pi$ (figure 1(f)). In these figures we can appreciate the half-circular pillar defect in the lower section of the disk, in this place never sit any vortices, because it acts like an anti-pinning defect. We can appreciate in

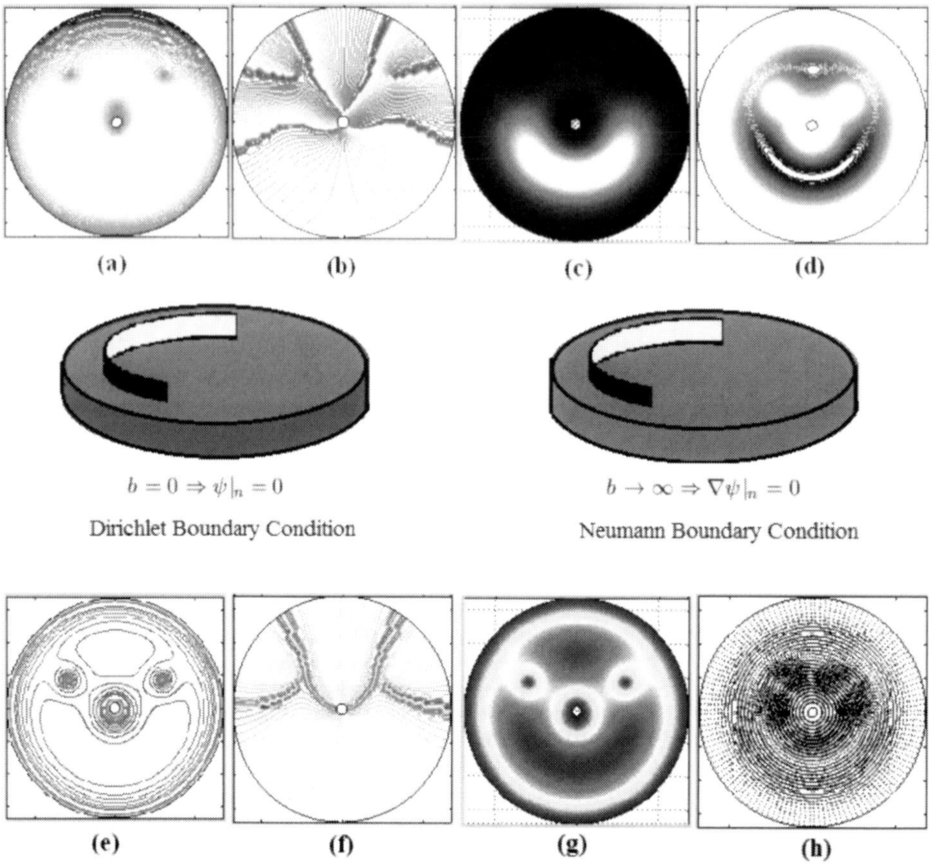

Figure 1. Square modulus of the order parameter $|\psi|^2$ (c, g) (in logarithmical scale (a, e)) and its phase $\Delta\Theta$ (b, f). Magnetic induction $\vec{\mathbf{h}}$ (d), and super-current $\vec{\mathbf{J}}$ (h), for a disk with a defect at $H_0 = 0.888$ with $N = 6$ vortices (up) and $H_0 = 0.652$ $N = 4$ (down), using the Dirichlet condition.

the table the Gibbs free energy G, the magnetization $-4\pi M$, vorticity $N\Phi_0$ (calculated at H_{01}), and the critical fields H_{01}, H_{02}, H_{03}, dependents of the chosen boundary condition. The growth of the magnetization and free energy with Dirichlet condition means that the less the metallic boundary is ($b \to \infty$), the more diamagnetic the material is. Also, when we take the condition $\psi = 0$ in the surface allow the superconductivity be destroyed and nucleated at lower magnetic fields.

Acknowledgements
The authors would like to thank Edson Sardella, Ely Dannier V-Niño and A. Barba for their very useful discussions.

References
[1] J Barba-Ortega, Edson Sardella, J Albino Aguiar 2011 *Supercond. Sci. Technol* **24** 015001
[2] J Barba-Ortega, Edson Sardella, J Albino Aguiar 2013 *Mod. Phys. Lett B* **27** 4 1350025
[3] R Geurts, M V Milosevic, J Albino Aguiar and F M Peeters 2013 *Phys. Rev* **87** 024501
[4] J Barba-Ortega, Edson Sardella, J Albino Aguiar 2013 *Physica C* **485** 107
[5] J Barba-Ortega, Edson Sardella, J Albino Aguiar, E H Brandt 2012 *Physica C* **479** 49

[6] V H Dao, L F Chibotaru, T Nishio and V V Moshchalkov 2011 *Phys. Rev. B* **83** 020503
[7] A Vagov, A A Shanenko, M V Milosevic, V M Axt and F M Peeters 2012 *Phys. Rev. B* **86** 144514
[8] W D Gropp, H G Kaper, G K Leaf, D M Levine, M Palumbo and V M Vinokur 1996 *J. Comput. Phys* **123** 254
[9] J Berger and J Rubinstein 2000 *Connectivity and Superconductivity* (Germany: Springer)
[10] J Berger and J Rubinstein 1999 *Phys. Rev. B* **59** 8896
[11] P G de Gennes 1999 *Superconductivity of Metals and Alloys* (Colorado: Westview Press)
[12] M Thinkham 1996 *Introduction to Superconductivity* (New York: McGraw Hill)
[13] J Barba-Ortega, Edson Sardella, J Albino Aguiar, F M Peeters 2013 *Physica C* **487** 47
[14] J Barba-Ortega, Edson Sardella, J Albino Aguiar 2013 *Physica C* **492** 1

Journal of Physics: Conference Series **466** (2013) 012028 doi:10.1088/1742-6596/466/1/012028

Effect of potential shape and height on the density of the impurity states in nanotube with two quantum wells

J D González[1,2], J Barba-Ortega[2] and F Rondano[1]
[1] Grupo Teoría de la Materia Condensada, Universidad del Magdalena, Santa Marta, Colombia.
[2] Departamento de Física, Facultad de Ciencias, Universidad Nacional de Colombia, Bogotá, Colombia.

E-mail: jedgonzalezac@unal.edu.co

Abstract. Recently heterostructures with two GaAs/GaAlAs quantum wells (QWs) in a section of the tube layer has been fabricated and optical properties of the embedded QWs has been studied. The ground state binding energy of an off-axis donor in a cylindrical nanotube, containing two GaAs/GaAlAs quantum wells (QWs) in the presence of a uniform magnetic field is calculated as a function of the donor location as well as the density of states. A trial function for describing the asymmetric electron charge distribution is taken as a product of the combination of 1s and 2px, y subband wave functions and an unknown function that depends only on electron-ion separation. We found that the increasing the magnetic field the increasing the binding energy while the impurity is located in the QW1, whereas the opposite occurs when the impurity is located in the QW2. Two peaks in the curves of the binding energy, as a function of the impurity position, are also found as well as in the density of impurity states.

1. Introduction

The multilayer structures including strained layers are useful to form nanotubes and nanocoils [1, 2]. By using lattice-mismatched epitaxial layers that rolled up when freed from the substrate due to the built-in strain and a micro-tube including two GaAs/GaAlAs quantum wells (QWs) located at positions with different types of strain was fabricated and its optical properties before and after the fabrication process were investigated by photoluminescence (PL) spectroscopy. Micro-tubes containing two GaAs/AlGaAs quantum wells (QWs) in a section of the tube layer, and also the optical properties of the embedded QWs have been studied [3]. In our previous work we show that the low lying *1s* and $2p_{x,y}$ subbands of the free electron in nanotube, become almost degenerated as the width of the repulsive core grows and the mixing of these subbands in the presence of the off-axis donor no longer should be depreciated [4].The purpose of this paper is to analyze the ground state energy of the off-axis neutral donor and the density of the impurity states in cylindrical nanotube with two GaAs/AlGaAs quantum wells with and without the presence of a magnetic field oriented along the growth axis.

2. Theory

We consider the problem of an off-axis neutral donor impurity in a cylindrical nanotube, which containing two $GaAs/Ga_{1-x}Al_xAs$ quantum wells (*QWs*) in the presence of an uniform magnetic field, as can see in figure1a. For modelling a nanotube containing two $GaAs/Ga_{1-x}Al_xAs$ quantum wells (*QWs*), we use the confinement potential given by the following expression:

$$V(\rho) = V_1\theta(-\rho,-R_1,W) + V_2\theta(\rho,R_2,W) - V_3\theta(\rho,R_3+W,W) + V_4\theta(\rho,R_4,W), \quad (1)$$

where:

$$\theta(z,z_0,W) = \begin{cases} 0 \quad ; & z < z_0 - W \\ \left[(z-z_0)/W^2 - 1\right]^2 ; & z - W \le z < z_0 \\ 1 ; & z \ge z_0 \end{cases} \quad (2)$$

is soft-edge version of the Heaviside function, R_i, V_i represent the radii and heights of the barriers, respectively; W is a parameter related to the width of the transition region (see figure 1 (b)). When the magnetic field is applied the probability is higher in the first well for both free electronic states. Similar behavior of the electronic wave function when a magnetic field is included in a superlattice in figure 2 in Ref. [6] with our figure 1 (c) was found. Additionally, this figure is in good agreement with figures 2 and 3 (a) for the calculation of the ground state energy as a function of the donor position.

The material parameters m^*, the conduction-band effective mass and ε, the dielectric constant of the semiconductor heterostructure are assumed to be uniform throughout the heterostructure. The values of the physical parameters pertaining to GaAs ($m^*=0.067\,m_0$, and $\varepsilon=12.53$, where m_0 is the free-electronmass) are used in our calculations.

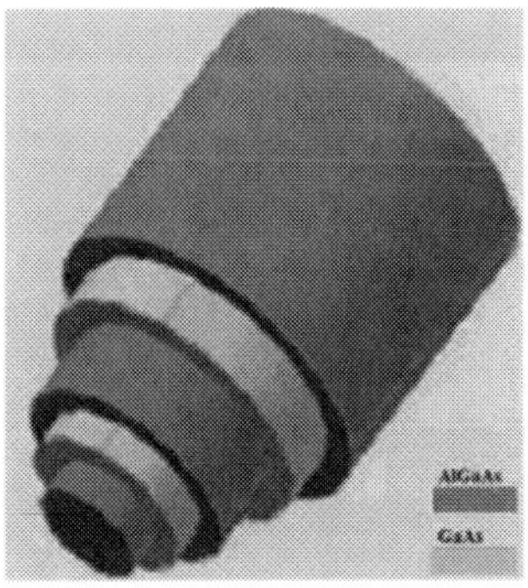

Figure 1. (a). Cylindrical nanotube, containing two GaAs/GaAlAs quantum wells (QWs).

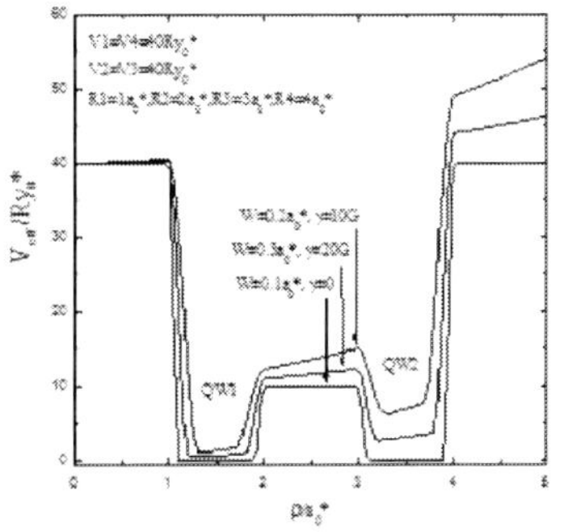

Figure 1. (b). Potential shape with different widths of the transition region.

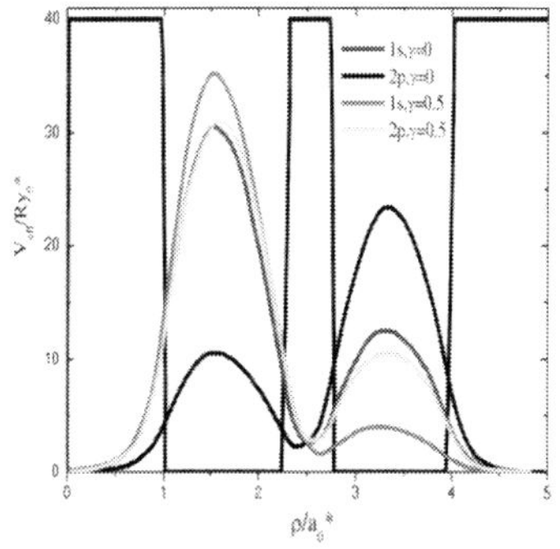

Figure 1. (c). Waves functions of the electron ground 1s and the first excited $2p_{x,y}$ for a neutral donor.

Within the effective-mass approach, the dimensionless Hamiltonian for the donor can be written as

$$H = H_0 - \frac{2}{|\vec{r}-\vec{\xi}|} \quad ; H_0 = -\nabla^2 + V(\rho) + \frac{1}{4}\gamma^2\rho^2 - i\gamma\frac{\partial}{\partial\varphi}, \quad (3)$$

Where $\vec{r}$ and $\vec{\xi}$ are used to designate the electron and ion positions, respectively and we have introduced the effective Bohr radius $a_0^* = \varepsilon\hbar/m^*e^2$ as the unit of length, the effective Rydberg $Ry_0^* = e^2/2a_0^*\varepsilon$ as the unit of energy and $\gamma = e\hbar B/2m^*cRy^*$ as the dimensionless unit of the magnetic-field strength corresponding to the first Landau level.

To calculate the donor ground state energy, we choose a trial function as the product of a linear combination of the waves functions of the electron ground, s and the first excited, $2p_{x,y}$ states, with an unknown isotropic function $\Phi(|\vec{r}-\vec{\xi}|)$ could be interpreted as a variational wave function:

$$\psi(\vec{r}) = \left[\alpha f_{1s} + \sqrt{1-\alpha^2} f_{2p}\right]\Phi\left(\left|\vec{r} - \vec{\zeta}\right|\right) \tag{4}$$

where α is a variational parameter $(-1 < \alpha < 1)$ which gives the grade of the mixing of the subbands [4]. For our model the wave equation for the free electron $H_0 f(\vec{r}) = E_0 f(\vec{r})$ is separable and the differential equation is solved numerically using the trigonometric sweep method [5]. Starting from the variational principle and using the method described in the papers [6] one can obtain the following Euler-Lagrange equation for the correlation function:

$$-\frac{1}{J(r)}\frac{d}{dr}\left[J(r)\frac{d\Phi(r)}{dr}\right] + \left[\tilde{E}(r) - \frac{2}{r}\right]\Phi = E(D^0)\Phi\,, \tag{5}$$

Where $E(D^0)$ represents the energy of the neutral donor, $J(r)$ is radial part of the Jacobian and $\tilde{E}(r)$ is the averaged free electron local energy. Finally, to define the donor energy we solve the wave equation (5) by using the trigonometric sweep method [5]. To calculate the density of the impurity states we assume that the circular cross section of the nanotube is not too small in order that we might treat the impurity positions as a continuous random variable. In this case the density of the impurity states $g(E)$ is [4]:

$$g(E_b) = 2\pi\xi_\rho(E_b)\left|dE_b(\xi_\rho)/d\xi_\rho\right|^{-1} \tag{6}$$

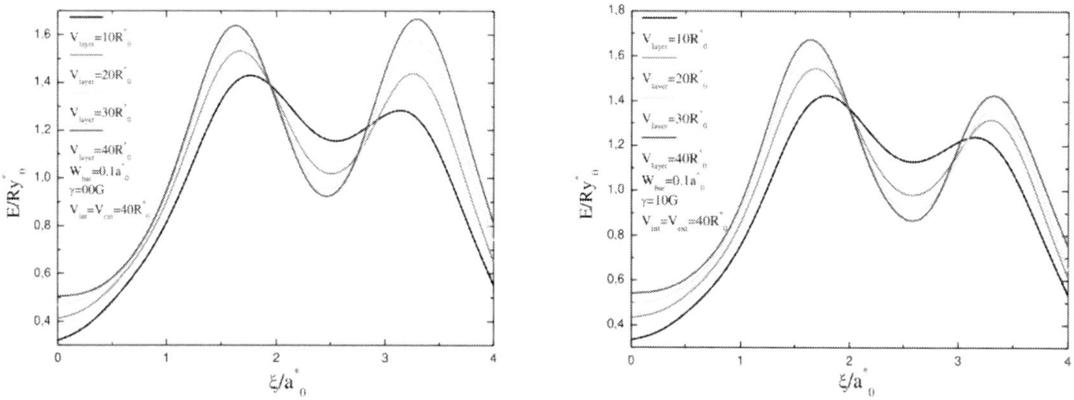

Figure 2. D^0 ground state binding energies as a function of the donor displacement ε_ρ from the axis of the cylindrical nanotube with two square quantum well with square potential, different values of central barrier V_{layer} and magnetic field.

Figure 2 shows a successive displacement of the curves peaks position to the right in the QW1 whereas the opposite occurs in QW2 with reducing V_{layer} due to the electron have high probabilities to penetrate the central potential barrier. While the minimum of the curves are found in the central barrier position when the height of the V_{layer} is increasing. The peak position coincides approximately with the midpoint between the interior and exterior radii for QW1 and QW2, $(R_1 + R_2)/2$ and $(R_3 + R_4)/2$respectively where the confinement is largest. The figure 3a shows the additional maximums of the energy due to the inclusion of the two square quantum well such behavior of the donor binding energies is also reflected in the curves of density of impurity states in nanotube. It seen from the figure 3b that the two well causes the appearance of the singularities near the right-side and left-side threshold of the curves of the density of the impurity states and it middle part. These singularities are due to the existence of one minimum and one maximum in the binding energies dependence where the derivate $dE(\xi_\rho)/d\xi_\rho$ is equal to zero.

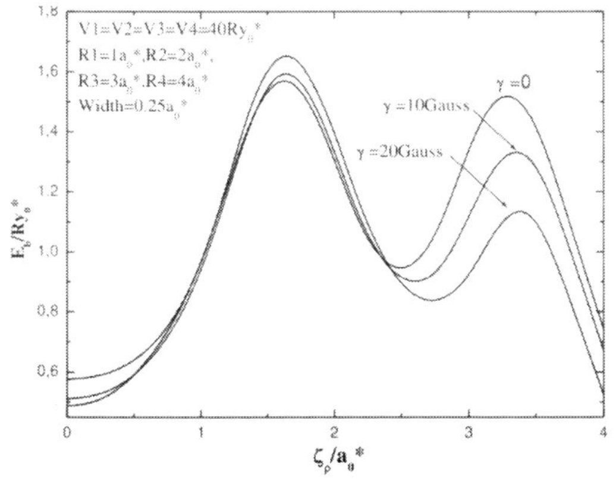

Figure 3. (a). D^0 ground state binding energies as a function of the donor displacement ξ_ρ from the axis of the cylindrical nanotube with different magnetic field values.

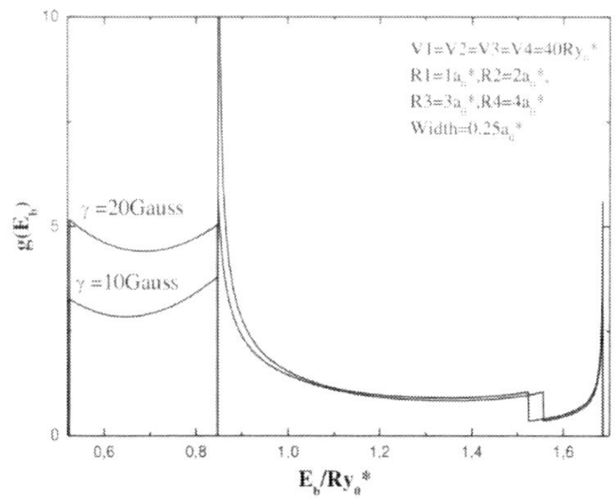

Figure 3. (b). Density of the D^0 impurity states as a function of the binding energy for different values of the magnetic field *B* applied.

3. Conclusions and recommendations

We have presented a simple method of calculation of the binding energy for the lowest state of the off-axis D^0 donors in a nanotube with two square wells in the presence of the magnetic field applied parallel to the axis, taking into account the effect of the mixing *s* y p_x subbands. We found that under external magnetic field applied parallel to the axis the binding energies of donors located close to the axis increases and located far from the axis decreases. Also we present new curves that show novels energy peaks and analyze the density of the D^0 impurity states for different magnetic fields.

Acknowledgement

This work was partially financed by the Universidad del Magdalena and the Colombian Agency COLCIENCIAS through doctoral scholarships 567.

References

[1] Prinz V Ya, Seleznev V A, Samoylov V A, Gutakovsky A K 1996 *Microelectron. Eng.* **30** 439
[2] Prinz V Ya, Seleznev V A, Gutakovsky A K, Chehovskiy A V, Preobrazenskii V V, Putyato M A, Nenasheva L A 1999 *Compound Semiconductors Institute of Physics Conference Series* **166**
[3] Kubota K, Vaccaroa P O, Ohtania N, Hiroseb Y, Hosodab M, Aida T 2002 *Physica E* **13** 313
[4] Betancur F J, Orozco E A, Gonzalez J D and Mikhailov I D 2005 *Phys. Stat. Sol.* **242** 1833
[5] F J Betancur, I D Mikhailov and L E Oliveira 1998 *J. Appl. Phys.* D **31** 3391
[6] I D Mikhailov, F J Betancur, R Escorcia and J Sierra-Ortega 2002 *Phys. Stat. Sol. (b)* **234** 590

Pillars and pinning centers in superconducting prisms

J Barba-Ortega[1], J Barón-Jaimez[2] and M R Joya[1]

[1] Departamento de Física, Universidad Nacional de Colombia, Bogotá, Colombia.
[2] Instituto de Investigaciones en Materiales, Universidad Autónoma de México, DF, México.

E-mail: jjbarbao@unal.edu.co

Abstract. We solve the Ginzburg-Landau equations for to study the vortex configurations in a superconducting prism with a square array of pillars or holes in the presence of a uniform applied magnetic field. The presence of the pillars (holes) changes the vortex structures in the superconducting prism considerably. We calculate magnetization, free energy and vorticity curves, which shows the transition between different vortex configurations as a function of magnetic fields.

1. Introduction

Superconducting samples with arrays of pinning or anti-pinning centers have received much attention over the last decade. Most interest has been focused on arrays of pillar and holes in mesoscopics squares, rectangles and disks [1, 2, 3]. The superconducting materials of high critical temperature have structural defects in a wide variety of shapes leading to complex dynamic and static processes. Therefore, it becomes of particular interest to simulate this class of defects for optimize the thermodynamical parameters for a given superconducting sample. Experimentally electron-beam technique allow one to design superconducting films with a large array of pillars of different shapes and sizes. These systems enable to study possible vortex configurations, to identify its stable and metastable configurations and consider the effect of the sample geometry on the final vortex state [4, 5]. In previous works we have used the Ginzburg Landau formalism to study the vortex matter in superconducting disks with different kinds of defects [6, 7, 8], in this work, we consider a superconducting prism with an array of superconducting pillars and holes.

2. Theoretical formalism

Let us consider a thin superconducting prism (square and circular) with a array of holes and pillars, immersed in an insulating medium in the presence of a perpendicular uniform magnetic field $\mathbf{H_e}$. Superconducting matter is described in the Ginzburg-Landau theory by the order parameter ψ and the potential vector $\mathbf{A}$, (for more details, see Ref. [9, 10, 11]) as:

$$\frac{\partial \psi}{\partial t} = -(i\nabla + \mathbf{A})^2\psi + \psi(1 - |\psi|^2) \quad (1)$$

$$\frac{\partial \mathbf{A}}{\partial t} = \mathrm{Re}\left[\bar{\psi}(-i\nabla - \mathbf{A})\psi\right] - \kappa^2 \nabla \times \nabla \times \mathbf{A} \quad (2)$$

The dynamical equations are complemented with the appropriate boundary conditions for the order parameter considering a superconducting/vacuum interphase: $(-i\nabla - \mathbf{A})\psi \cdot \mathbf{n} = 0$, $\hat{n}$ is the unity vector perpendicular to the surface of the superconductor [12, 13].

3. Results and discussion

The parameters used in our numerical simulations for a prism of pure Al were $\xi(0) = 1.2\ \mu m$, $T_c = 1.7K$, $T = 0.85$, the square and circular area of the prism was $S = 64.0\xi^2(0) = \pi R^2$. The area of the defects in the square is $d = 9.0\xi^2(0)$, the disk has point-like defects. G is the free Gibbs energy, M is the magnetization, N is vorticity in units of fluxoid Φ_0 and H_{01} is the first entry of vortices magnetic field, H_{02} is the second critical field and H_{03} is the upper magnetic field. The figures 1 (b, c, d, f, g) shown the contour plot of the order parameter for

Table 1. Thermodynamics values for a superconducting cylinder and square prism at H_{01}.

Sample	$-G(H_{01})$	$-4\pi M(H_{01})$	$N\Phi_0(H_{01})$	H_{01}	H_{02}	H_{03}
Square prism with defects	0.615	0.620	4.0	0.650	0.920	2.229
Square prims without defects	0.735	0.852	4.0	0.880	1.020	2.550
Cylinder with defects	0.725	0.722	3.0	1.275	1.45	2.750
Cylinder without defects	0.725	0.722	3.0	1.275	1.45	2.750

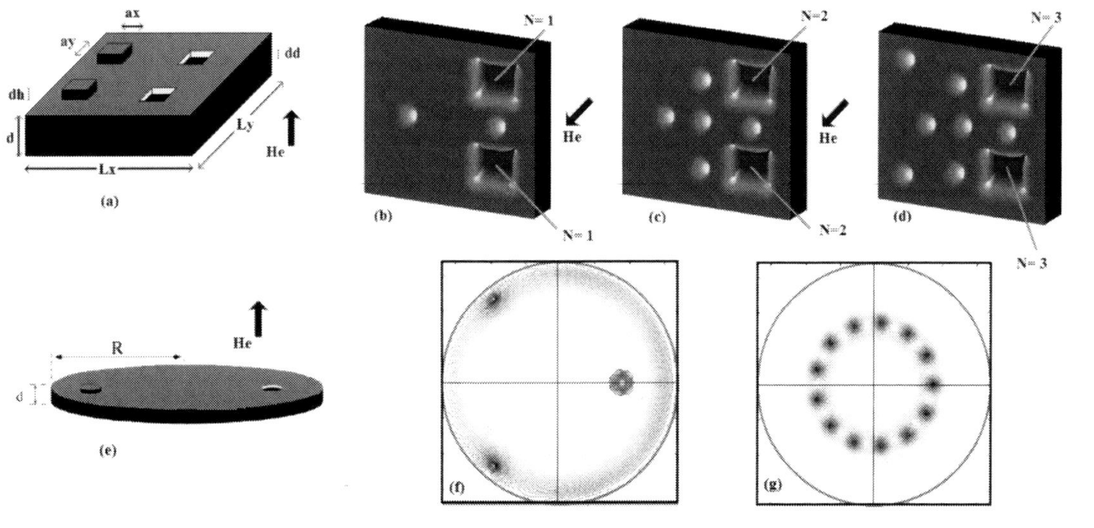

Figure 1. layout of the samples (a) Square prims with two square dots and two square anti-dots (e) Cylinder with one dot and one anti-dot. (b, c, d, f, g) Square modulus of the order parameter $|\psi|^2$ for at H_{01} with $N = 4, 9, 13$ vortices (square) and $N = 3, 13$ (cylinder).

several vorticities at magnetic field at H_{01} with $N = 4, 9, 13$ vortices (square) and $N = 3, 13$ (cylinder); $N = 2, 4, 7$ vortices remains in the superconductor region, while one, two and three vortices sit in the each hole of the square. Although they are not visible in the contour plot of the magnitude of order parameter, there is a change in the phase around the hole equal to $\Delta\Theta = 2N\pi$, these defects broken the typical symmetry of the vortex square configuration. In these figures we can appreciate the point like defect (pillar) in the cylinder, in this place never sit any vortices, because it acts like an anti-pinning defect. We can see in the table the Gibbs free energy G, the magnetization $-4\pi M$, vorticity $N\Phi_0$ (calculated at H_{01}), and the critical fields H_{01}, H_{02}, H_{03}, dependents of the size of the defects, for a point like defect (small defects) the magnetic fields are independent of its nature and size. The growth of the magnetization and free energy with big defects means that more diamagnetic the material is. Also, this defects allow the superconductivity be destroyed and nucleated at lower magnetic fields.

Acknowledgements
The authors would like to thank Edson Sardella, Ely D. Valbuena, A. Barba and M. A. Escobar for their very useful discussions.

References

[1] V Moshchalkov, M Baert, V Metlushko, E Rosseel, M Van Bael, K Temst, R Jonckheere and Y Bruynseraede 1996 *Phys. Rev. B* **54** 7385
[2] G R Berdiyorov, M V Milosevic and F M Peeters 2006 *Phys. Rev. Lett* **96** 207001
[3] G R Berdiyorov, V Misko, M V Milosevic, W Escoffier, I V Grigorieva and F M Peeters 2008 *Phys. Rev* **77** 024526
[4] I V Grigorieva, W Escoffier, J Richardson, L Y Vinnikov, S Dubonos and V Oboznov 2006 *Phys. Rev. Lett* **96** 077005
[5] I V Grigorieva, W Escoffier, V R Misko, B J Baelus, F M Peeters, L Y Vinnikov and S V Dubonos 2007 *Phys. Rev. Lett* **99** 147003
[6] J Barba-Ortega, Edson Sardella, J Albino Aguiar 2013 *Physica C* **485** 107
[7] J Barba-Ortega, Edson Sardella, J Albino Aguiar, E H Brandt 2012 *Physica C* **479** 49
[8] J Barba-Ortega, Edson Sardella, J Albino Aguiar, *Modern Physics Letter B* 2013 **27** 4 1350025
[9] W D Gropp, H G Kaper, G K Leaf, D M Levine, M Palumbo and V M Vinokur 1996 *J. Comput. Phys* **123** 254
[10] J Berger and J Rubinstein 2000 *Connectivity and Superconductivity* (Germany: Springer)
[11] Edson Sardella and E H Brandt 2010 *Supercond. Sci. Technol* **23** 025015
[12] P G de Gennes 1999 *Superconductivity of Metals and Alloys* (Colorado: Westview Press)
[13] M Thinkham 1996 *Introduction to Superconductivity* (New York: McGraw Hill)

Journal of Physics: Conference Series **466** (2013) 012030 doi:10.1088/1742-6596/466/1/012030

Density of states for a light-hole exciton in a microtube of GaAs/AlGaAs with two quantum well and different potential shape: theoretical model

A D Barrios[1] **J. Barba-Ortega**[2] **and J D González**[1,2]
[1] Grupo Teoría de la Materia Condensada, Universidad del Magdalena, Santa Marta, Colombia.
[2] Departamento de Física, Universidad Nacional de Colombia, Bogotá, Colombia.

E-mail: jedgonzalezac@unal.edu.co

Abstract. We consider a simple method for calculating the ground-state energy of light hole exciton and density of states confined in a microtube of GaAs/AlGaAs recently fabricated and studied experimentally, with a double quantum well and large radius of curvature. The exciton trial function is taken as a product of the ground state wave functions of both the unbound electron and hole in the heterostructure, with an arbitrary correlation function that depends only on electron-hole separation. A renormalized Schrödinger equation for the correlation function is derived and coincides with the corresponding equation for a hydrogen atom in an effective and space-isotropic homogeneous. The binding energy of the ground state to an exciton in this heterostructure, the contribution to the energy given by the sublevels and the density of states is determined as a function of the width of the well, the aluminium concentration and confinement potential profile is obtained by solving the equation calculated by the variational model proposed.

1. Introduction

It is well known that nanoscale technology has made possible the fabrication of low-dimensional heterostructures with controlled thickness and relatively sharp interfaces, where the excitons remain present even at room temperature because of the quantum confinement increases highly the electron–hole attraction [1]. A great deal of attention has been devoted to experimental studies of excitons in heterostructures based on III–V semiconductor [2–8] particularly quantum wells (QWs), superlattices (SLs), quantum-well wires (QWWs), quantum dots (QDs), quantum rings (QRs) and more recently microtube [9]. The multilayer structures including strained layers are useful to form nanotubes and nanocoils [10]. By using lattice-mismatched epitaxial layers that rolled up when freed from the substrate [11-13] due to the built-in strain and a micro-tube including two GaAs/GaAlAs quantum wells (QWs) [14, 15] located at positions with different types of strain was fabricated and its optical properties before and after the fabrication process were investigated by photoluminescence (PL) spectroscopy [16]. The purpose of this paper is to analyze the density of the states of a light-hole exciton as a function of the width of quantum wells taking into account the ground state energy and the contribution given by sublevels in a microtube with two GaAs/AlGaAs quantum wells.

2. Theory

We propose a simple variational procedure in which the problem of a light-hole exciton in a microtube with two $GaAs/Ga_{1-x}Al_xAs$ quantum wells (QWs) is considered as we can see in the figure 1 (a). For modelling a microtube, we use the confinement potential figure 1 (b) given by the following expression:

$$V(z) = V_0\theta(-z, -L/2, W) + V_0\theta(z, L/2, W) \qquad (1)$$

$$\theta(z, z_0, W) = \begin{cases} 0; & z < z_0 - W \\ [(z - z_0)^2/W^2 - 1]^2; & z_0 - W \le z < z_0 \\ 1; & z \ge z_0 \end{cases} \tag{2}$$

The wave function of the free electron and light-hole is Shown in the figure 1 (c). The strong confinement due to the narrow quantum well used, shows a mixing of the wave functions and strong penetration in the bulk.

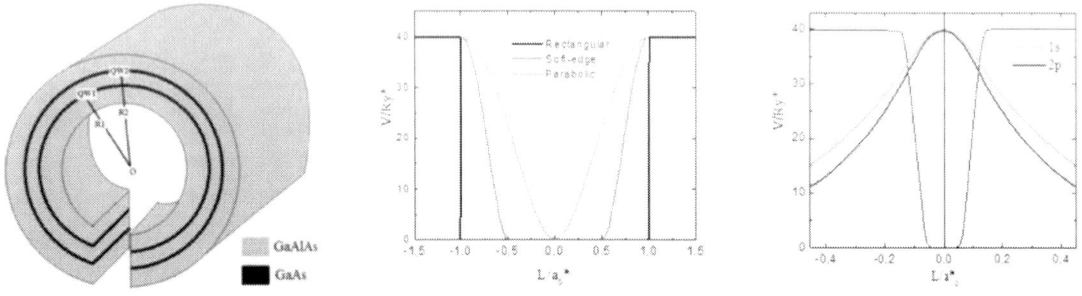

(a) **(b)** **(c)**

Figure 1. (a) Cylindrical microtube, containing two GaAs/GaAlAs quantum wells (QWs). (b) Potential shape with different widths of the transition region. (c) Waves functions of the electron ground 1s and the first excited 2p for a light-hole exciton.

In this work, we consider a single exciton in semiconductor heterostructures within the effective mass approximation by using an spherical hole mass approximation and constant physical parameters throughout the heterostructures [10]. In this case the characteristic energy of an exciton can consequently be defined through the effective Rydberg $R_y^* = e^2\ /2\varepsilon\, a_0^*$, and its characteristic length through the effective Bohr radius, $a_0^* = \varepsilon\hbar/\mu e^2$, with $\mu = m_e^*\, m_h^*/m_e^* + m_h^*$ being the reduced exciton effective mass. Thus the dimensionless Hamiltonian for the electron-hole pair confined in a heterostructure in the presence of an external field may be written as:

$$H(\boldsymbol{r}_e, \boldsymbol{r}_h, \tau) = H_e(\boldsymbol{r}_e) + H_h(\boldsymbol{r}_h) + V(\boldsymbol{r}_e, \boldsymbol{r}_h) - 2\tau/r_{eh} \quad (\tau = 0{,}1) \tag{3}$$

$$H_e = -\eta_e \nabla_e^2 + V_e(\boldsymbol{r}_e);\ H_h = -\eta_h \nabla_h^2 + V_h(\boldsymbol{r}_h) \tag{4}$$

Where H_e and H_h describe the free motion of electron and hole confined in the heterostructure respectively, $V(\boldsymbol{r}_e, \boldsymbol{r}_h)$ is a potential confinement, $-2/r_{eh}$ is the energy of the electron-hole interaction , $\boldsymbol{r}_e$ and $\boldsymbol{r}_e$ are electron and hole position vectors, and $r_{eh} = |\boldsymbol{r}_e - \boldsymbol{r}_h|$ is the electron-hole separation. The parameter τ in equation (1) is equal to 1 for exciton and to 0 for uncorrelated electron-hole pair, and the parameters $\eta_e = \mu/m_e^*$ and $\eta_h = \mu/m_h^*$ in equation (4) are characteristics of the corresponding dimensionless effective masses of the hole in units of the reduced exciton effective mass μ respectively. For an uncoupled electron-hole pair ($\tau = 0$), the Hamiltonian is separable, on the contrary for exciton ($\tau = 1$), and the Schrödinger equation can only be solved by using some approximation methods. Assuming that the center of mass motion is affected by the electron-hole interaction to a considerably smaller degree than the relative motion, we choose the ground state exciton trial function in the form:

$$\Psi(\boldsymbol{r}_e, \boldsymbol{r}_e) = f_0(\boldsymbol{r}_e, \boldsymbol{r}_h)\Phi(r_{eh});\ r_{eh} = \sqrt{(x_e - x_h)^2 + (y_e - y_h)^2 + (z_e - z_h)^2} \tag{5}$$

where $f_0(\boldsymbol{r}_e, \boldsymbol{r}_h)$ is the product of the one-particle electron and hole ground state wave functions, whereas $\Phi(r_{eh})$ is a variational function that describes the intrinsic properties of the exciton and depends only on the electron-hole separation. The ground state energy of the exciton is found by minimizing the functional, after some algebraic manipulations we can obtain the following variation problem:

$$F[\Phi] = \int_0^{\infty} \left\{ J_0(r) \left[\left(\frac{d\Phi(r)}{dr} \right)^2 + \left(E_0 - E_{ex} - \frac{2}{r} \right) \right] \Phi^2(r) \right\} dr \rightarrow min \qquad (6)$$

with $E_0 = E_e + E_h$ being the ground state energy of the uncoupled electron-hole pair, $r = r_{eh}$ and

$$J_0(r) = \int d\boldsymbol{r}_e \int f_0^2(\boldsymbol{r}_e, \boldsymbol{r}_h) \delta(|\boldsymbol{r}_e - \boldsymbol{r}_h| - r) d\boldsymbol{r}_h \qquad (7)$$

The minimization of the functional (6) with respect to Φ and its first derivative yields the Euler-Lagrange equation

$$\frac{1}{J_0(r)} \frac{d}{dr} J_0(r) \frac{d\Phi(r)}{dr} - \frac{2}{r} \Phi(r) = [E_{ex} - E_0] \Phi(r) \qquad (8)$$

The binding energy of the exciton, $E_b = E_0 - E_{ex}$, is then obtained by solving numerically equation (8). In the figure 2 the inclusion of the two quantum wells is reflected in the curves of density of excitonics states in microtube. The energy of sublevels are obtained including the contribution the rotational kinetic energy which is added to the exciton energy and gap, may be written as:

$$E_i(m) = \Delta_i + \frac{m^2}{R_i^2} \qquad (9)$$

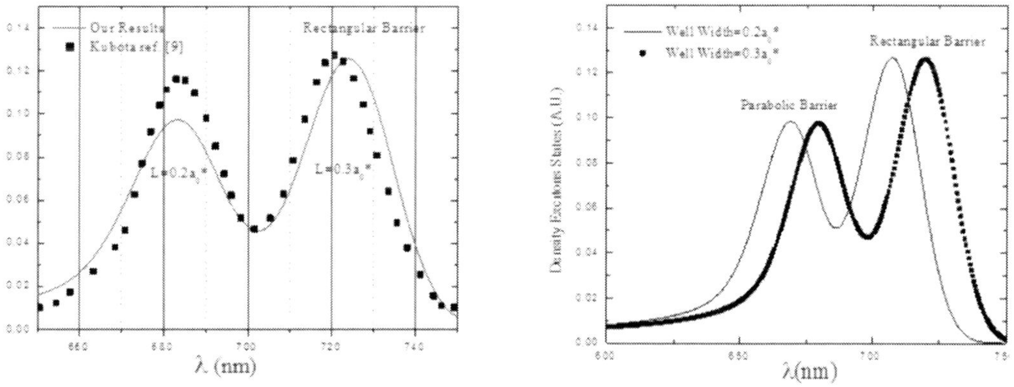

(a) **(b)**

Figure 2. Shift in the peak position in the state density curve in function of the wavelength for different shapes of the confinement potential.

Where $i = 1,2$, Δ_i is the exciton energy plus the gap energy, m magnetic quantum number and R_i radii from center of microtube since QW1 or QW2 figure1 (b). figure 2. Shows the density of states for a light-hole exciton in a microtube, with radii $R_1 = 2.5a_0^*$ $R_2 = 5.7a_0^*$ corresponding to the width of the well $L_1 = 3nm = 0.3a_0^*$ y $L_2 = 2nm = 0.2a_0^*$ respectively. This graph shows that the behavior of excitonic states density is similar to that shown by photoluminescence curve ref. [9], high coincidence of peaks corresponding to wavelengths in both curves is shown. Moreover, the difference in the height of the peaks in figure 2 (a) and (b). is due to the choice in the location of wells QW2 and QW1, for radii R_1 and R_2 in the microtube. The higher peaks in the QW's is located in the position corresponding to the value of R with smaller radius of curvature. figure1 (b) shows additional confinement produced by interdiffusion $Ga_{1-x}Al_xAs$ material into the well in GaAs, while figure 2 (b) shows the achieved energy due to the interdiffusion, in each well and hence in the position of each peak. We see higher energy values are achieved by using smaller well widths. Additionally, the parabolic profile provides a higher potential than the rectangular profile confinement. Figure 2 (a) The curve measured experimentally by photoluminescence [9] is compared with the energy density of states of a light hole exciton with square well profile and radius of the first well QW1, $R_1 = 2.5a_0^*$ and radius of the second well QW2, $R_2 = 5.7a_0^*$.

3. Conclusions

We have presented a simple method of calculation of the binding energy for the lowest state of the light-hole in a microtube with two wells with different potential shapes, taking into account the sublevels due to the rotational kinetic energy. We found that ours calculus are in a good agree with experimental case. Also we present new curves that show novels density of states peaks and different configuration of a microtube can be analyzed.

Acknowledgement

This work was partially financed by the Universidad del Magdalena and the Colombian Agency COLCIENCIAS through doctoral scholarships 567.

References

[1] D A Miller 1998 *Optical Switching in Low-Dimensional Systems* (New York: Plenum)
[2] R L Greene, K K Bajaj, D E Phelps 1984 *Phys. Rev.* B **29** 1807
[3] J Cen, K K Bajaj 1992 *Phys. Rev.* B **46** 15280
[4] L Gang, S V Branis, K K Bajaj 1995 *J. Appl. Phys.* **77** 1097
[5] S Le Goff, B Stebe 1993 *Phys. Rev.* B **47** 1383
[6] T Garm 1996 *J. Condens. Matter.* **8** 5725
[7] J M Ferreyra, C R Proetto 1998 *Phys. Rev.* B **57** 9061
[8] S Glutsch, F Bechstedt, W Wegscheider, F Schedelbeck 1997 *Phys. Rev.* B **56** 4108
[9] K Kubota, Vaccaroa P O, Ohtania N, Hiroseb Y, Hosodab M, Aida T 2002 *Physica E* **13** 313
[10] V Prinz, Ya, Seleznev V A, Samoylov V A, Gutakovsky A K 1996 *Microelectron. Eng.* **30** 439
[11] I S Chun, K Bassett, A Challa and X Li 2010 *Applied Physics Letters* **96** 251106
[12] R D Rodriguez 2012 *Nanoscale Research Letters* **7** 594
[13] F Balhorn, Mansfeld S, Krohn A, Topp J, Hansen W, Heitmann D, Mendach S 2010 *Phys Rev Lett.* **104** 037205
[14] G L Miranda, M E Mora-Ramos, C A Duque 2013 *Physica B* **409** 78
[15] N Syrbu, A Dorogan, N Dragutan, T Vieru, V Ursaki 2011 *Physica E* **44** 20
[16] M Nakayama, T Hirao, T Hasegawa 2010 *Physica E* **42** 2644
[17] Y F Lao 2010 *Applied Physics Letters* **97** 091104

Comparison of mechanical behavior between implant-simulated bone tissue and implant-jaw bone tissue interfaces based on Pull Out testing

C Lopez[1], J C Muñoz[1] and J C Pinillos[1]
[1] Interfaz research group, Industrial University of Santander, Bucaramanga, Colombia.

E-mail: clara.lopez@correo.uis.edu.co

Abstract. The main purpose of this research was to achieve a better understanding of the relationship within the mechanical properties of human cadaver jaw bone with kind D2 density regarding a substitute polymer to simulate bone tissue, proposed by the ASTM, to evaluate orthopedic implants. However, despite the existence of several densities of foams and his mechanical characterization has been classified into different degrees of tissue densities to simulate cancellous and cortical bone, the value of the densities are different contrasted with the densities of bone tissue, making difficult to establish direct relationship about mechanical behavior between the polymer and the bone material, and therefore no clear criteria known for choosing the polymeric foam which describes the mechanical behavior of tissue for a specific or particular study. To understand such behavior from bone tissue regarding the polymer samples, on this research was a dental implant inserted into the samples, and subjected to destructive Pull Out test according to ASTM F543The Pull Out strength was compared between implant-jawbone and implant-rigid polyurethane foam interfaces. Thus, the test pieces with mechanical behavior similar to bone tissue, enabling an approximation to choose degree appropriate of polymer to replace the bone tissue in future trials biomechanical.

1. Introduction

The scientific community has developed studies based on design, materials and biomechanical analyzes of dental implants for replacement of dental roots. These biomechanical studies are performed to validate the design of the implant. Nevertheless, the jaw bone is formed by cortical and trabecular bone with a percentage of occurrence that depends of their biomechanical function on the jaw arch [1], for example, in the molar region the percentage of cancellous bone is higher than cortical bone tissue, so that its density is of type D2 and D3, according to the clinical classification dental [2].

However, due to the destructive nature of the biomechanical testing, ASTM suggest that destructive testing with implants for any product that requiring the use of bone tissue, the bone must be replaced by a polymer specimen developed to simulate bone tissue (Rigid Foam polyurethane) [3].

Nevertheless although it is known that bone density is related to the mechanical strength [1, 2], there is no coincidence between the mechanical and physical properties of bone tissue with respect to the polymer, whereby it is not possible to establish selection criteria describing the mechanical behavior equivalent to bone tissue. However, a way to know the interface strength between implant and bone specimen's polymer is through destructive test Pull Out [4], which was proposed in this paper to establish by comparison, the similar mechanical behavior between bone tissue specimens and specimens of polyurethane.

Journal of Physics: Conference Series **466** (2013) 012031 doi:10.1088/1742-6596/466/1/012031

From mechanical testing set out to establish a relationship of the kind of density jaw bone sample with the bone specimens simulated. In the following sections describe the materials used in mechanical testing, methodology, results and discussion of results.

2. Methods and materials

The first step was design a dental implant to molar region, with 12 mm length and 5 mm diameter, threaded body type HB, ASTM [4]. The implant was made of Ti6Al4V alloy [5] by specialized surgical company. Subsequently, were used three specimens of human cadaver mandible in the molar region, the material was supplied and prepared in the legal medicine laboratory of the Faculty of Health Industrial University of Santander; in turn, the bone specimens were characterized by technique imagenological [6], so the tissue density of molar region was identified as D2 type 2gr/cc3 [7]. Polymer samples were chosen to simulate trabecular bone density type using G12, G15, G20, G20/12, cortical bone density and type G30, G30/12, G40/10, G40/12, ASTM F1839-08 [3]; for each specimen degree were performed three experiments, data on polymer characterization of the samples were provided by the manufacturer Sawbone® [8] (see table 1).

The implants were inserted in the samples through a digital torque wrench [9]. Subsequently the samples were subjected to Pull Out destructive test, based on ASTM F543 used to measure the magnitude of the insertion force generated by the contact surface at the bone-implant [10], through the application of an axial load until remove the implant from the specimen [11]. The tests were made in the laboratory GIMAT; the assemblies were tested in the universal testing machine from Tinius Olsen 50KN load [12]. In figure 1 left, it sees the image of the specimen located on the universal testing machine Tinius Olsen. Figure 2 corresponding to the central image that shows the e dental implant machined in Ti6Al4V inserted into a Sawbone® specimen's; the image was taken under microscope KH-7700 HIROX [13], in laboratory GIMAT research group. Finally in Figure 3 shows the removed implant bone interface like result of the Pull Out test.

Figure 1. Preparing the test on Tinius Olsen Machine.

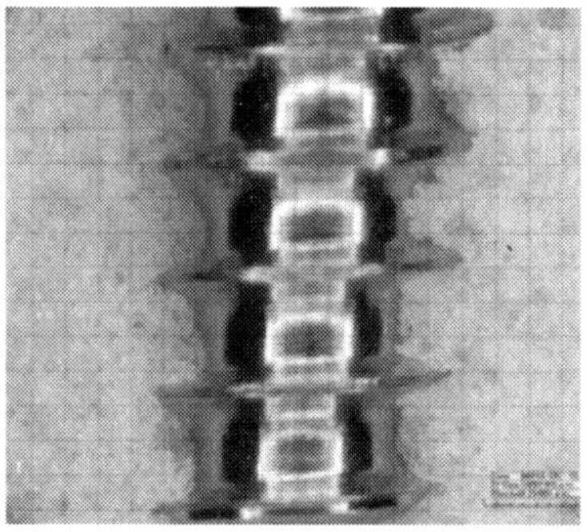

Figure 2. Sectional view of the bone-implant.

Figure 3. Implant removal interface testing machine.

3. Results

Once were made the assay, from these were obtained the extraction load values and displacement data on the implant interface, from this data was calculated the elastic limit stress, strain, Young's modulus E and Max. Stress, as shown in table 1. Likewise, the mechanical behavior of the specimens was plotted in figure 4 by the graphic Force / Displacement. Specimens like G12 at G20, with supposed mechanical strength similar to graphic "bone", were again plotted for comparative purposes, like seen in figure 5. Thus, were identified the specimens with mechanical behavior similar to bone.

Table 1. Physical and mechanical properties characterization based on Pull Out test.

Specimen	Density (gr/cm3)	Strain (elastic limit) (mm/mm)	Load/Elastic Limit (N)	Sy[a] (MPa)	E[b] (Gpa)	Strain (mm/mm)	Max. Load (N)	Max. Stress (MPa)
Bone	2.0	0.005	190.8	9.71	1240.8	0.037	336.0	17.1
G 12	0.2	0.021	106.1	5.40	331.3	0.046	151.6	7.7
G 15	0.2	0.011	140.5	7.15	476.1	0.030	184.0	9.3
G 20	0.3	0.019	223.9	11.40	729.9	0.040	323.8	16.4
G 20/12	0.2	0.012	133.2	6.78	475.2	0.070	321.6	16.3
G 30	0.4	0.003	48.0	2.44	351.0	0.100	734.2	37.3
G 30/12	0.2	0.006	67.5	3.43	256.7	0.170	489.0	24.9
G 40/10	0.1	0.007	39.2	1.99	218.0	0.100	129.2	6.5
G 40/12	0.2	0.009	22.4	1.14	82.9	0.130	270.0	13.7

[a] Stress Yield
[b] Young´s Modulus

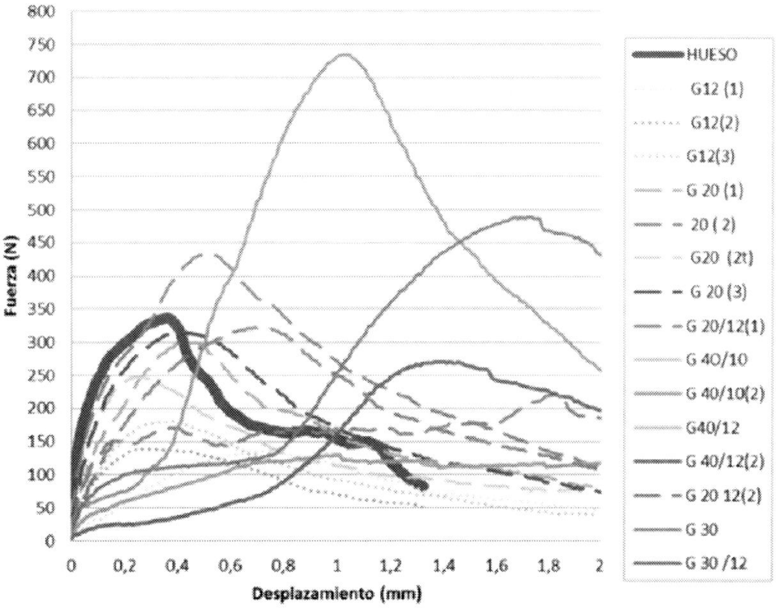

Figure 4. Outcomes assay according to Pull Outtest, Graphic force/ displacement of total.

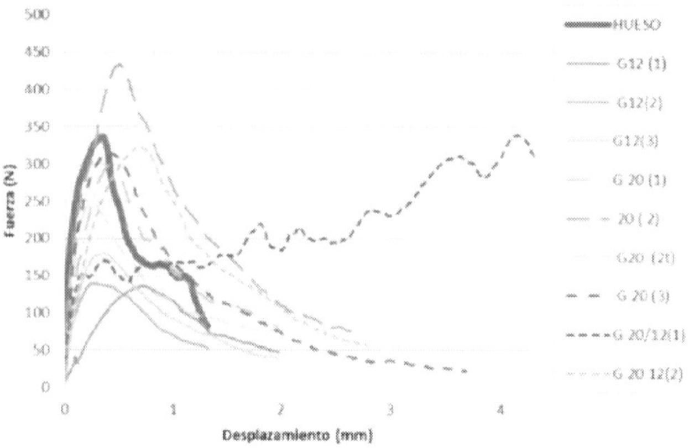

Figure 5. Comparison between mechanical behavior of bone specimen and the samples G12.

4. Discusion of outcomes

The pull out tests made in this study allowed to know the mechanical behavior of specimens implant interfaces by which it was possible to establish the mechanical behavior of the implant extraction resistance at the interface between the specimens G 20, regarding bone specimen. Thus, it was established for these experiments that the G20 grade specimen has a relation with the mechanical properties of bone mandibular molar region D2 density type, so it was possible to select this grade of specimen for future experiments based on experimental and simulation tests.

Further testing is required for the samples with G20 density combined with lower density than grade G20, to establish a suitable selection because there was no enough consensus regarding the mechanical behavior in each of the tests.

It is recommended to extend the study to bone densities D1, D3 and D4 Kind according classification proposed by Misch, thus establishing a direct link between the sawbone® specimen and the bone tissue, because these outcomes will be useful in selecting suitable polymeric material to replace living tissues in others physical and mechanical destructive test.

Acknowledgements

The authors would appreciation to the VIE Vice-chancellorship research and outreach Industrial University of Santander, for their financial support in the framework of the "Research for the design of an orthopaedic implant to the extent based on imaging techniques and biomechanical evaluation of the implant bone interface by the finite element method, for the company specialized surgical ", developed for the consolidation of research groups with funded.

References

[1] Turkyilmaz I and Mc Glumphy E A 2008 *BMC Oral Health* **8** 32
[2] Misch C E, Qu Z and Bidez M W 1999 *J. Oral and Maxillofacial Surgery* **57** 700
[3] American Society for testing and Materials ASTM F1839-08 2009 (State United: ASTM)
[4] American Society for Testing and Materials ASTM F 543–07´1 2009 (State Unite: ASTM)
[5] Niinomi M. 2003 *Science and Technology of Advanced Materials* **4** 445
[6] Taylor W, Roland E, Ploeg D, Klabunde R, Warnera M and Hobathod M 2002 *Journal of Biomec.* **35** 767
[7] Homolka P, Beer A, Birkfellner R, Nowotn W, Gahleitner A and Bergmann H 2002 *Radiology* **224** 247
[8] Crawford L A, Powell E S and Trail I 2012 *J of Hand Surgery rest.* **37** 255
[9] Eliasa C, Rocha F, Nascimento A and Coelho P 2012 *J. of the Mech. Behavior of Biomed. Mat.* **16** 169
[10] Ferraz O N, Shimano A C, Marc J M, Penazzo L C and Candido dos Reis A 2013 *Inter. J. of Dentistry* **1**
[11] Berzins A, Shah B, Weinans H and Sumner D R 1997 *Journal of Biomedical Materials Research* **34** 337
[12] Testing 2008 *Plastics, Additives and Compounding* **10** 37
[13] Valiavalappila S and Harinipriya S 2012 *Synthetic Metals* **162** 2027

Journal of Physics: Conference Series **466** (2013) 012032 doi:10.1088/1742-6596/466/1/012032

Optimization of the bamboo guadua angustifolia kunth in the elaboration of glued laminated elements for constructive use

G A Díaz[1,2], R A Cruz[2] and A M Chávez[2]
[1] Industrial Design School, Industrial University of Santander, Bucaramanga, Colombia.
[2] Materials and Construction Structures Research Group, Civil Engineering School, Industrial University of Santander, Bucaramanga, Colombia.

E-mail: gadiazr@uis.edu.co

Abstract. Bamboo is considered one of the best timber resources in the world because for its mechanical properties and high sustainability; this research aims to improve the mechanical properties of the laminated glued bamboo Guadua Angustifolia Kunth (GAK) for use as structural elements, starting from de very manufacture process; this is important because it is possible to observe variations in the flexural strength and the elastic modulus in GAK samples taken from different heights and thickness of the culm.

In order to analyze the influence of these final mechanical properties variations in the laminated, the height of the culm where samples are extracted (cepa, basa and sobrebasa) it is taken as a variable from where different types of laminated were manufactured, seeking to make optimal the configuration based in the transversal section area and the material strength.

Three assemblies were designed varying the overlap of the adhesion lines and it concluded that the highest strength average values were obtained in the laminated composites manufactured with samples taken from the bottom of the culm (basa), which is possible because in these elements there are less adhesion lines than the other ones (middle, top and mixed) or the better matching of themselves.

1. Introduction

The bamboo - Guadua is considered one of the wooden species with the best physical-mechanical properties and high sustainability. In addition, is known like one of the biggest bamboos of the world, in which it is possible to find 30 meters of highness and 22 centimetres of diameter figure 1 [1].

The transversal section of Guadua Angustifolia Kunth (GAK) presents differences in the radial direction, where the parenchyma cells, vascular bundles and fibres are located.

The fibre concentration varies according the height of the stem and in the radial direction from de outer to the inner diameter. As a result, the mechanical properties of the GAK vary significantly according to the height section and the thickness of the specimen [2].

This study seeks to improve the mechanical properties of GAK derivate products, specifically the glued laminated composites used for structural applications by analysing the manufacturing process.

Figure 1. Bamboo Plant type: Guadua Angustifolia Kunth.

2. Materials and Methods

To the laminates elaboration, stems 75 cm long were cut off from three different cross section heights (bottom, middle and top); and afterwards, the sheets were extracted and machined in order to conform the glued laminated composites.

At the end, the composites had 40*40 [mm] on the cross section according to the NTC301 standard (Colombian Technical standard), figure 2 and 3.

Figure 2. Initial cuts in the Bamboo GAK.

Figure 3. Bamboo GAK glued laminates.

2.1. The structural glued laminated manufacturing process: important aspects

One of the most important aspects in the elaboration of these composites is the quality of the interfacial or adhesion line, this depends of the surface quality of the raw material and the adhesive selected; about this issue, prior studies have shown that the best adhesive is the based in melamine resin because of his mechanical resistance and durability against humidity, however, in this study it was taken as a research topic only the mechanical behaviour, so a polyvinyl based adhesive was used in this case [3].

Another study aspect is how to achieve an improvement in the use of the raw material, making it more efficient. For the elaboration of the sheets that conforms the laminates near the 30% to 50% of the total stem it was wasted; however, if the right technologies are proposing, it is possible to reduce that percentages [4].

The raw material was obtained from Induguadua S.A., this company has the support of the autonomous regional corporation of Quindío Department (CRQ), who is in charge of grant permissions and environmental licences in this geographical area.

2.2. *Mechanic characterization*

The mechanical properties of the laminated composites were obtained by flexural testing according to NTC 663 standard (Colombian technical standard). The equipment used for the test was a Multitest System MTS 50, with a load cell of 500 Kn. The maximum values obtained ranged between 69 and 141 Kg-f/cm^2 (see figure 4).

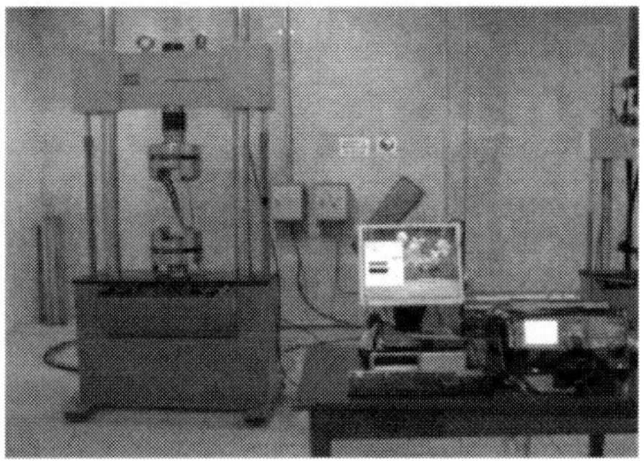

Figure 4. Initial cuts in the Bamboo GAK.

Figure 5. Bamboo GAK glued laminates.

There was 30 samples tested, manufactured with different types of sheets specimens in order to ensure the independence of the results; the relative humidity was 12% and the stems were mature (between 4 to 6 years) (see figure 5).

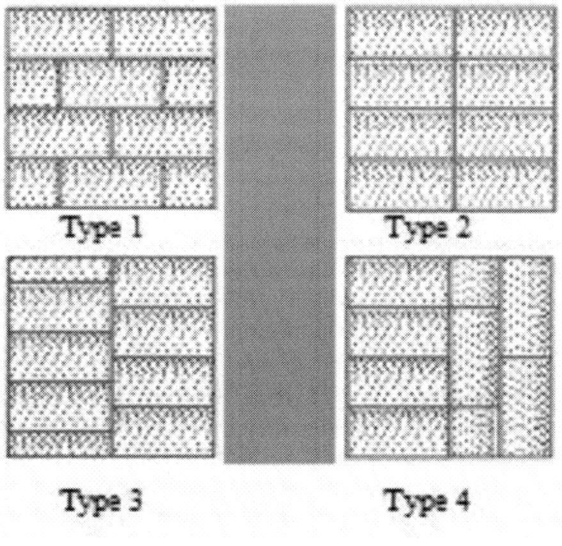

Figure 6. Four different types of special orientation sheets for the laminates.

3. Results

The composites exhibited densities between 0,743 and 0,731 [g/cm^3], it was observed that there aren't exist significant differences between the laminates made with bottom sheets (type 1) and the other combinations; the average values of the densities in the combined configuration composites (types 3 and 4) from middle and top sheets was 0,722 [g/cm^3], close to the densities of middle and top.

The final stress values show that the best configurations were the bottoms sheets based ones, located between 71,4 Mpa end 56,6 Mpa.

The stresses were calculated from the stress-strain diagram, and then it was made a comparison table about the mechanical properties founded, in which it is possible to look that the weight/stiffness relation is favorable if it is compare with others construction materials (see figure 7).

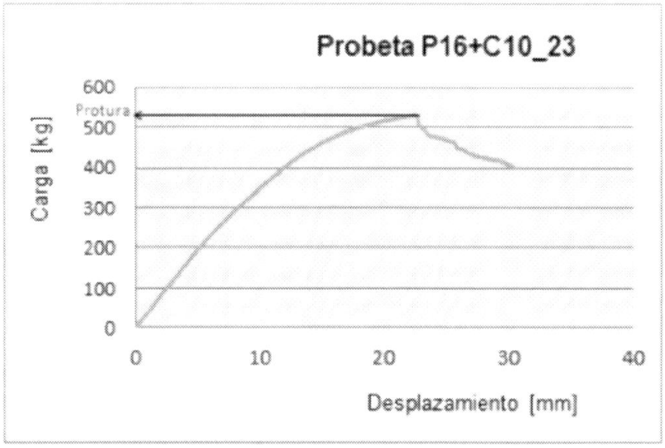

Figure 7. Load vs. displacement curve of a laminated type 1 based on bottom sheets sample.

In general, the resistance values of the whole laminates set had a similar mechanical behavior; the difference between each type of laminate configuration was the quantity of adhesion lines required for the manufacturing process.

4. Conclusions

In the bamboo glued laminated manufacturing process, the most important issues are the material mechanical properties of the raw material, the sheets geometry and its spatial orientation.

The laminates manufactured with bottom sheets showed the best mechanical properties in the flexural stresses, with maximum values 54 kg-f/cm^2.

The sheets with the biggest fibers concentration are the middle and top ones, however the laminates with the best mechanical behavior were the composites made with bottom sheets (type 1), this it is possible perhaps this type of laminates need less raw material and therefore less adhesion lines.

Acknowledgements

Authors wish to acknowledge the assistance from the civil engineering students crew, especially Fitzgerald Lozano and Ana Maria Chavez, and besides the support of the technical staff of the construction materials characterization research group in the Industrial University of Santander.

References

[1] Londoño X 1998 *J. Am. Bam. Soc.* **12** 37
[2] Lo T, Cui H Z, Leung H C 2004 *J Mat. Lett.* **58** 2595
[3] Gonzalez H A, Hellwing S, Montoya J A 2009 *J Fac. Cie. Agrop.* **7** 1
[4] Lopez L F, Correal J F 2009 *J Mad. Cie. Tec.* **11** 171

Journal of Physics: Conference Series 466 (2013) 012033 doi:10.1088/1742-6596/466/1/012033

Characterization and design of asphalt mixtures with asphaltites from Boyacá for use in low traffic volume roads

R Manrique-Espindola[1]
[1] Universidad Pedagógica y Tecnológica de Colombia - UPTC, Tunja, Colombia.

E-mail: rmanriques@gmail.com

Abstract. The high availability of asphaltites in Boyacá and their low cost make this material a viable alternative for low traffic road paving; nevertheless, the traditional way in which this material is used generates, in cases, pavements with deficient behavior. This investigation, presents the results of the mixture design using asphaltites from the municipality of Pesca-Boyacá as well as coarse and fine aggregates produced in the region, 70-80 asphalt cement and slow-break asphalt emulsion. Working formulas for dense mixing in hot and cold and particularly MDF-2 and MDC-2 are presented from the characterization information; as benchmarks to define technical viability for use in low- traffic volume roads, according to NT1 regulation from INVIAS. The mixture design was performed according to the procedures defined in the RAMCODES and MARSHALL methodologies.

1. Introduction

The deficiency in vial infrastructure is a factor that reduces competitiveness in a region, affecting its development and limiting its inhabitants regarding the possibility of improving their life conditions, this is why it is necessary to seek for viable alternatives for cost reduction in paving works, thus obtaining higher coverage of roads in optimal conditions for communities and ultimately, contribute to the economic and social development of the region.

Keeping in mind the proper use of asphaltites form the municipality of Pesca in the department of Boyacá, this research, undertaken by the research group GRINFAVIAL from the "Masters in Engineering with emphasis in Infrastructure" at the Universidad Pedagógica y Tecnológica de Colombia, intends to perform the characterization of the exploited material in the quarries of Santa Teresa and La Emilia in this municipality, in order to identify the physical and chemical characteristics of this material and thus being able to propose an asphaltic mixtures design for use in low-traffic volume roads.

2. Asphaltic mixtures

Asphaltic mixtures in hot are defined as the combination of a hydro-carbonated binding agent, aggregates such as mineral dust and, eventually, additives, in such a way that all the particles in the aggregate are thoroughly covered by a uniform film of binding agent. The asphaltic mixture in cold is a mixture of mineral aggregates with or without mineral fillings, along with emulsified or diluted asphalt; the whole process is carried out at environment temperature, (Jimenez, 2009).

In the asphalt mix design include the following materials:

- The asphaltic cement 70-80 for MDC-2 and for MDF-2 type is slow-break cationic emulsion.

- Natural asphalt (asphaltite) from the defined quarries in the municipality of Pesca - Boyacá.
- The coarse aggregates required for granulometric stabilization are obtained as a product of triturating.

The design of hot and cold dense mixtures for roads with low-traffic volume will be obtained using the protocols defined by the MARSHALL and RAMCODES methods; this is done in order to obtain the working formula with optimal percentage of asphalt and that also adjusts to the quality parameters defined by the INVIAS-07.

3. Material characterization

It is in this chapter that we show the results of characterization of materials used for asphaltic mixtures with natural asphalt form the Santa Teresa and La Emilia quarries in the municipality of Pesca. This way we define the investigation in four work groups, two per quarry, in such a way that a group performs the design of the dense mixture in cold MDF-2 and the other one the dense mixture in hot MDC-2 for each quarry.

3.1. Asphaltite characterization

The information obtained from the laboratory shows a fine, improperly gradated material, composed by sandstone impregnated in asphalt, divided in 96% and 4% proportions approximately; this material is fit for the design of asphaltic mixtures, given its physical properties and according to the specifications defined by INVIAS art. 400-07 (see table 1).

Table 1. Asphaltite characterization results.

Standard INVIAS-07	Test	Standard Value	MDC-2 Santa Teresa	MDF-2 Santa Teresa	MDC-2 La Emilia	MDF-2 La Emilia
E 133-07	Sand equivalent Soils and Fine aggregates (%)	50% Min	66	67	91	95
E 222-07	Specific gravity of fine aggregates (Gsb)	% Absorptión	2.44	2.5	2.35	2.23
E 222-07	Specific gravity of fine aggregates (Gsb-sss)	% Absorptión	2.53	2.57	2.44	2.292
E 222-07	Specific gravity of fine aggregates (Gsa)	% Absorptión	2.69	2.7	2.58	2.386
E 222-07	Absorption %	0.0365	0.04	2.86	3.83	2.86
E 707-07	Specific gravity of fine aggregates (Gb) – Lixiviado	N.A	1.09	0	0.98	1.038
E 732-07	Extracción Cualitativa de Asfalto (%)	N.A	4.05	4	3.89	3.9
E 125-07	Liquid limit	≤40	17.18	15.27	16.43	18.11
E 126-07	Plastic limit	4 a 9	N.P	N.P	N.P	N.P
E 706-07	Penetration	200 - 250	224.78	224.7	235	277
E 723-07	Distillation	N.A	N.P	N.P	N.P.	N.P.
E 709-07	Flash Point y Cleveland Open Cup °C	≥200°C	93	97.28	188	92

Experiments performed for chemical characterization of asphaltites were carried out at the carbon laboratory in the UPTC. In this characterization the analysis was performed on asphalt-impregnated sandstone and the bituminous component extracted from asphaltites form the Santa Teresa and La Emilia quarries in Pesca- Boyacá. Likewise, the chemical analysis performed on the sand stones hows an ash-composed material in more than 90% in both quarries; it also shows a material that does not present dilatation or contraction at temperatures up to 520°C. Also, it does not present plastic behavior, maintaining its constant fluidity (30000 dial/min) up to a temperature of 500°C.

3.2. Aggregate characterization

Stony aggregates, bearing in mind that average measured parameters show results that fit the defined limits of technical specifications art. 400-07 INVIAS. All of the previous is focused on composition, cleansing, durability, geometry and specific gravity (see table 2).

Table 2.Stony aggregate characterization.

Standard INVIAS-07	Test	Standard Value	MDC-2 Santa Teresa	MDF-2 Santa Teresa	MDC-2 La Emilia	MDF-2 La Emilia
E 227-07	Percentage of fractured faces (2 faces)	60 % min.	100 %	100 %	100 %	100 %
E 230-07	Flattening index	Max 35	27.80 %	37.91 %	29.41 %	37.10 %
E 230-07	Elongation index	Max 35	26.27 %	54.51 %	25.98 %	56.11 %
E 222-07	Specific gravity of thick aggregates (Gsb)	N.A.	2.53	2.60	2.57	2.60
E 223-07	Specific gravity of thick aggregates (Gsb-sss)	N.A.	2577.00	2.63	2.62	2.63
E 223-07	Specific gravity of thick aggregates (Gsa)	N.A.	2.65	2.69	2.71	2.69
E 223-07	Absorption %	1.69	1.69	1.24	1.91	1.69
E 218-07	The wear resistance of machine of angeles aggregates	30 %	21.43 %	21.22 %	23.08 %	21.60 %
E 125-07	Liquid limit	Min 40	16.43	16.05	17.72	19.67
E 126-07	Plastic limit		13.51	12.75	14.99	18.11
E 125-07	Plastic index	4 a 9	2.92	3.30	2.73	1.56
E 220-07	Aggregates health against the action of sodium sulfate	Max 12	7.62 %	8.00 %	7.62 %	2.60 %

3.3. Bitumen characterization

Verified properties in bituminous products adjust to the requirements of technical specifications, defined by INVIAS art. 400-07 (see table 3).

Table 3. Bitumen Characterization results.

Bitumen	Standard INVIAS-07	Test	Standard Value	MDC-2 Santa Teresa	MDC-2 La Emilia
Asphalt Cement	E 706 – 07	Penetration (1/10) mm	70 -80	78.67	80.67
	E 723 – 07	Destilation (%)	min 50	81	83
	E 707 – 07	Specific Gravity Bulk Gbulk	N.A	1.131	1.069
	E 716/717-07	Saybolt Furol viscosity (SSF)	60-120 SSF	60.5	61.01
	E 712 -07	Softening point ring and bal (°C)	30 - 200 °C	59.1	50.9
Asphalt Emulsion	E 763-07	Saybolt Furol viscosity (SSF)	0 - 200	25	24
	E 761 -07	Water content by volume (%)	Max 43	40	39
	E 767-07	Loading particles		positive	Positive
	E 762-07	Distillation residues (%)	Min 57	60.0	62
		P.H	Max 6	2.81	2.82
		Retained sieve No 20	Max 0.1	0.0135	0.014

4. Asphaltic mixture design

Starting from material characterization, granulometry was adjusted according to the INVIAS regulation for MDC-2 mixture and NT-1 traffic, thus defining the coarse aggregate and asphaltite percentages for the different groups. (Art 400-07).

For each proposed combination, we determined the optimal asphalt percentage using the Marshall method, according to Illinois methodology. Stability, flow, air voids, mineral aggregate voids and asphalt-filled voids were determined in this briquette; this was made with the intention of verifying with the values established by the INVIAS regulation. Values for stability and flow as main quality check parameters for this asphaltic mixture obtained for the MARSHALL and RAMCODES methods.

Once the optimal asphalt percentage, which included asphaltite, was obtained for each of the asphaltic mixtures and once the compliance of regulations was verified, the definite work formulas and the summary of design for mixtures in each group was summarized in table 4.

Table 4. Coarse aggregate and asphaltite percentages in propsed mixtures.

	Agregado	MDC-2 Santa Teresa	MDF-2 Santa Teresa	MDF-2 la Emilia
Upper band	Coarse aggregate %	57	58	65
	Asphaltite %	43	42	35
Central strip	Coarse aggregate %	61	52	70
	Asphaltite %	39	48	30
Lower band	Coarse aggregate %	64	48	87
	Asphaltite %	36	52	10

Mixtures made with an addition of natural asphalt as fine mineral aggregate and asphaltic contribution, show results on control parameters defined in the MARSHALL method, found within the parameter limit (INVIAS art 450-07) thus showing the effect of asphaltite presence, mainly in flow, since this parameter varies with asphaltite content in the work formula.

Table 5. Verification of experiment results in asphaltic mixtures according to specifications inv-450-07.

Standard INVIAS-07	Criteria/Test	Standard Value (NT-1)	MDC-2 Santa Teresa	MDF-2 Santa Teresa	MDF-2 la Emilia
	Asphalt percentage		6.5	5.5	6.6
	Coars aggregate percentage		61	52	70
	Asphaltite percentage		39	48	30
	Band		Central	Central	Central
E 227-07	Compaction (blows/face)	50	50	50	50
E 748-07	Minimum estability (Kg)	Min 500	603	570	641.5
E 748-07	Flow (mm)	2-4	3.74	6.5	3.861
E 799-07	Air gaps (Vv) Rolling	3-5	4	4	3.6
E 799-07	Gaps in mineral aggregates (VAM)	15	15.4	15.35	15.6
E 799-07	Voids filled with asphalt (VFA)	65 - 80	74.1	73.8	77.2

5. Conclusions

The information obtained from the characterization experiments in the materials that conform the asphaltic mixture with asphaltite allow us to state that natural asphalt is a material that shows acceptable conditions for cold and hot asphaltic mixture elaboration, with the intention of use in asphalt roads NT1, since the compliance with quality standards was verified, according to regulations from the INVIAS.

Parameters determined for quality control in the asphaltic mixture (stability, flow, air voids, voids in mineral aggregates and asphalt filled voids) comply with values established in the INVIAS regulation, both in the MARSHALL and RAMCODES method, which allows us to state the technical viability of natural asphalt usage from Santa Teresa and la Emilia quarries in Pesca- Boyacá for the design and construction of MDC-2 and MDF-2 rolling layers for NT1 traffic roads.

Resulting work formulas for the designed mixtures allow us to ensure that the range defined by the granulometric strip are covered by the technical characteristics observed in the experimented briquettes, thus providing reliability regarding the unavoidable variations of size in aggregate particles and asphaltite.

The results obtained allow us to conclude that asphaltite is a material that proportionally affects quality in a mixture, since its higher volume percentage in the mixture produces higher values in flow, which can generate inadequate behavior from the mixture in the presence of heavy vehicles, slow traffic or high temperatures.

References

[1] Camacho C 1989 *Estudio de Asfaltos Naturales de Boyacá como Capa de Rodadura* (Tunja: Universidad Pedagógica y Tecnológica de Colombia)

[2] Chavarro H y Gómez F 1980 *Utilización de una Mezcla Asfalto-Asfaltita en Pavimentos para Vías* (Bucaramanga: Universidad Industrial de Santander)

[3] Fundación Laboratorio Nacional de Vialidad 2003 *Diseño de Mezclas Asfálticas en Frío Empleando Emulsiones*

[4] Guevara M, Méndez J y Pimentel J *2010 Diseño de Mezclas Asfálticas Densas en Frío Basado en el Método Marshall Modificado de la Universidad de Illinois* (Salvador: Universidad de Salvador)

[5] Instituto de Desarrollo Urbano de Bogotá 2006 *Especificaciones Técnicas para el Diseño y la Construcción de Capas Estructurales de Pavimentos con Asfaltos Naturales en Frío*

[6] Instituto Nacional de Vías 2007 *Especificaciones de Materiales para la Construcción de Carreteras*

[7] Macucci F y Pinto A 2000 *Estudio Técnico de Asfalto Natural* (Tunja: PAVICOL)

[8] Mayorga A y Obregón R 1990 *Utilización de Asfaltitas en Pavimentos* (Bucaramanga: Universidad Industrial de Santander)

[9] Pérez G 2008 *Materiales para Ingeniería Apuntes de clase* (Tunja: Universidad Pedagógica y Tecnológica de Colombia Facultad de Ingeniería)

[10] Rematoso C y Villabona H 1979 *Estudio para el Mejoramiento de Capas de Pavimentos con Asfalto Natural de Pesca* (Bucaramanga: Universidad Industrial de Santander)

[11] Sanchez F J, Garnica P, Gómez J A, Pérez N 2003 *RAMCODES: Metodología Racional para el Análisis de Densificación y Resistencia de Geomateriales Compactados* (Queretaro: Instituto mexicano del Transporte)

[12] Sánchez F J 2004 *Aplicación de RAMCODES en el diseño y control de compactación de suelos 4as Jornadas Internacionales del Asfalto Corasfaltos* (Colombia: Cartagena)

[13] Sandoval J A 2007 *Estudio geológico y Caracterización de las Asfaltitas de la Vereda Paloblanco en el Municipio de Vélez - Santander* (Sogamoso: Universidad Pedagógica y Tecnológica de Colombia)

Journal of Physics: Conference Series **466** (2013) 012034 doi:10.1088/1742-6596/466/1/012034

Analysis of the use of coal tar as a binder in bituminous mixtures, using Marshall and Ramcodes methodologies

R Ochoa-Díaz[1]
[1] Universidad Pedagógica y Tecnológica de Colombia - UPTC, Tunja, Colombia.

E-mail: ricardo.ochoa@uptc.edu.co

Abstract. This paper presents an alternative use of coal tar, a by-product of the steel industry, given the problems of accumulation and negative environmental impact. Therefore, it is necessary to analyze the incorporation of coal tar as a binder in paving mixtures. First, this paper presents the origin, description of the main characteristics, and properties of tar. Then, this paper evaluates the mix of coal tar by means of the RAMCODES and Marshall methodologies to determine its resistance. The results of the tests explain the physical and mechanical properties of the mix. Taking into account the results of both methods, this paper makes a comparison to determine the suitability of the RAMCODES methodology in the mix design. Finally, it analyzes the alternatives to coal tar that can be used as binders in bituminous mixes for pavement and the advantages of their uses under some specific conditions.

1. Introduction

Since ancient times, man has continually developed activities that allow him to subsist and evolve. At the beginning, these activities were limited to consuming the animals and plants that nature offered, without producing any alterations in nature since the residues generated by these activities were mainly organic. With the development of industrial activities, however, mankind has produced an increasing number of inorganic residues that are not biodegradable. Governments and institutions are dedicating resources to the development and promotion of technologies that allow these inorganic residues and materials to be used constructively, thereby reducing the environmental and economic consequences that they cause [1].

The construction of highways is an activity that is generally considered to be an important use of natural resources in which, specifically, the creation of bituminous mixtures used in the construction of wearing courses requires the use of large quantities of stone aggregates [2].

In order to promote environmentally sustainable practices, this study aims to contribute to the development of technologies that utilize the residues and byproducts of the steel industry in order to improve the environmental conditions in the areas of influence of these companies. The objective is to analyze and evaluate the possibility of substituting part of the natural aggregates and the asphalt that are traditionally used in the manufacturing of bituminous mixtures for residues from the steel industry like granulated slag and coal tar.

2. Materials

2.1. *Coal tar as a binder*

Tar is a semisolid bituminous product or liquid that is obtained as a residue of the distillation, in the absence of air, of organic substances that posses volatile materials, like carbon or wood. It is a binder in the cold, water resistant, insoluble in lubricating oils and petroleum derived combustibles, soluble in ether, benzene, carbon disulfide, chloroform, and quinoline. Tar is formed from aliphatic and naphthalene compounds that contain hydrogen, oxygen, nitrogen, and sulphur.

2.2. Aggregates
Crushed gravel was used as a thick aggregate, which was subjected to the following tests: Los Angeles abrasion coefficient, percent of fractured face, elongation index and flattening index. Sand was used as a fine aggregate. Portland cement and fine sand that passed through a #200 sieve were used as mineral filler, both of which were tested to determine their specific gravity. All of the tests were conducted according to the INVÍAS testing guidelines for road materials [3].

2.3. Granulated slag as a fine aggregate
Granulated slag is a residue that comes from the blast furnace in the manufacturing of iron in integrated steel plants. The blast furnace is a tall, vertical oven whose main function is to produce pig iron of a chemical constant composition. The elements that are released from a blast furnace are: blast furnace gas, blast furnace flue dust, liquid slag and pig iron. The slag is released from the blast furnace in liquid form at 1400 °C and is pushed forward with a high pressure water jet and disintegrated into small particles to form granulated slag [4, 5].

3. Methodology
The methodology utilized in this investigation can be divided into three stages. First, tests were conducted to study the characteristics of the materials to be used in the design of the bituminous mixtures. Second, tests were conducted to obtain the working formula of the bituminous mixture using the RAMCODES and Marshall methodologies [6-8]. Third, the results obtained using these two different methodologies were compared as well as the benefits of utilizing tar through the obtainment of the dynamic module of the mix that offers the best mechanical characteristics.

4. Experimental design
This study attempted to implement the mix for wearing courses and the mix chosen for the development of this study is a type 2 heat-dense mixture (MDC-2), article 450-07 INVIAS [9]. With the idea of conducting a detailed analysis of the behavior of the slag and the coal tar in the mix, fifteen types of mixes were designed with the following variables: the fine aggregates (slag (E) and sand (A)) and mineral filler (cement (C) and fine sand (A)). The grain sizes of these materials were adjusted inside the granulometric band of the MDC-2 mix; they were placed in the center, on the fine side and on the thick side of the granulometric band. These were adjusted with the objective of evaluating the incidence of the grain size in the behavior of the mix.

5. Results

5.1. Physical and mechanical properties
The results obtained provide evidence for the possibility of creating bituminous mixtures utilizing coal tar as a binder. The best results were obtained when cement was used as filler in the mixture. There was no significant difference detected between mixes that used slag and mixes that used sand as fine aggregates, but in the mixtures with slag it was necessary to use a higher percentage of binder.

The prepared mixture that presented the best characteristics and that fit the standards demanded by the specifications for transit level NT-2 is the M-GAC-14 mix, which is presented below in table 1. This mix was tested to obtain the characteristic dynamic.

5.2. Dynamic properties
This test was performed with the N.A.T. (Nottingham Asphalt Tester) following the guidelines prEN 12697-26 Method IT-CY to an uncertainty of ±1.146% for a level of confidence of 95%.

The dynamic modules were made at a frequency of 10 Hz and at three different temperatures: 5°C, 25°C and 40°C. The results are displayed in figure 1.

Table 1. Results mix m-gac-14.

Characteristic	Specification Nt2	Results test	
		Marshall	Ramcodes
Minimum stability (Kg)	750	792.0	755.0
Flow (mm)	2 – 4	3.9	4.1
Air voids in the mix (Va) %	3 – 5	4.0	3.9
Voids in aggregate (VMA) %	15	15.2	15.8
Voids filled with binder (VFA) %	65 – 78	73.9	75.4

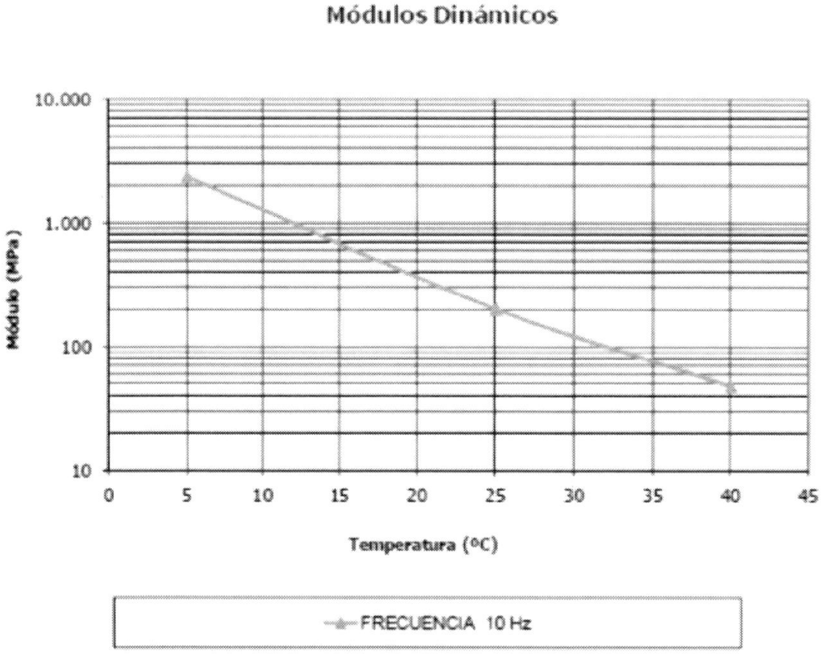

Figure 1. Dynamic module mix M-GAC-14.

6. Conclusions

This study aims to open an investigation to evaluate the behavior of bituminous mixtures for pavement using different types of binders other than asphalt.

This study does not aim to replace asphalt in pavement mixtures; the intention is to look for alternatives that allow manufacturers to select the product that is best suited to each individual project.

RAMCODES is a useful methodology for the design, production, and quality control of bituminous mixtures because it guarantees the completion of the volumetric parameters demanded by the specifications. This methodology also saves time, resources, and money, since it only requires three briquettes in comparison with the 15 briquettes required by the Marshall test.

The incorporation of the cement as a filler played an important role in the behavior of the mixture because it improved the properties of the binder and improved the adhesion of the aggregates, which permitted better stability and flow responses.

Keeping the thermal susceptibility of coal tar and consequently bituminous mixtures in mind, it was determined that these types of mixtures can be used in cold and temperate climates. Below these conditions, the behavior of the mixture will be acceptable according to the action of the load imposed by transit.

If some mixes did not meet the requirements demanded by INVIAS for wearing courses, they could be used as base or sub-base layers improvers.

References

[1] Belmonte Sánchez A 2009 *Análisis de la Reutilización de Residuos Procedentes de la Industria de Silestone en la Fabricación de Mezclas Bituminosas* (España: Universidad de Granada)
[2] Carmen R, Germán M, Luis B, Fernando M 2012 *Warm mix Asphalt: an Overview Journal of Cleaner Production* **22** 206
[3] Instituto Nacional de Vías 2007 *Normas de Ensayos de Materiales para Carreteras 1 y 2* (Bogotá: INVIAS)
[4] Alfonso G 2002 *Proceso Siderúrgico Planta Belencito* (Colombia: Acerías Paz del Río S.A.)
[5] Pérez S 2008 *Evaluación de la Escoria de Horno como Agregado en Mezclas Asfálticas* (Guatemala: Universidad de San Carlos de Guatemala)
[6] Sánchez F 2002 *Metodología Racional para el Análisis de Densificación y Resistencia de Geomateriales Compactados* (México: Publicación Técnica 200 ed. Sanfandila, Querétaro)
[7] Delgado H 2006 *Influencia de la Granulometría en las Propiedades Volumétricas de la Mezcla Asfáltica* (México: Publicación Técnica 299 ed. Sanfandila, Querétaro)
[8] Sánchez F 2008 *Manual de Aplicación Ramcodes* (Venezuela: Solestudios C.A.)
[9] Instituto Nacional de Vías 2007 *Especificaciones Generales de Construcción de Carreteras* (Bogotá: INVIAS)

Determination of stress concentration factors on flat plates of structural steel

A Santos[1]
[1] Grupo de Investigación en Desarrollo Tecnológico, Mecatrónica y Agroindustria – GIDETECHMA, Facultad de Ingeniería Mecánica, Universidad Pontificia Bolivariana seccional Bucaramanga, Floridablanca, Colombia.

E-mail: alfonso.santos@upb.edu.co

Abstract. The aim of this study was to determine stress concentration factors using finite element software; within the most outstanding results is ANSYS software validation for the determination of concentration factors in flat plates with a central hole subjected to axial load; Since it could be statistically demonstrated that there aren´t significant differences between the theoretical data and the data obtained through simulation. Were also obtained plots of stress concentration factors for flat plates with three holes, under the same load; it is noted that if the holes are transversely located, the concentration factor decreases to the extent that the relationship P3 / d increases, but to ensure this correlation it should be prevented that the hole diameter and the hole spacing exceeds a limit value, since if this occurs the point of maximum stress changes the inner to the outer area of the hole because the separation between the hole and the edge of the plate decreases significantly. If the arrangement of the holes is longitudinal it can be seen that the concentration factor decreases to the extent that the ratio d / b increases and additionally the magnitude of the factor is unaffected by the distance between centers of the holes and are quite similar to present a single flat plate with central hole.

1. Introduction

Abrupt changes originated from irregularities in the distribution of stresses are known as stress concentrators; these are presented for all types of stress, axial, bending or shear in the presence of fillets, holes, grooves, keyways, splines, tool marks or accidental scrapes. The inclusions or defects within the material over the surface also serve as "stress risers". The first mathematical study on stress concentration was published shortly after 1900, with the aim of working with other very simple different cases, experimental methods were developed to measure local efforts. In recent years they have started using computer simulations based on finite elements.

The information available in the literature is quite limited and is restricted to simple conditions of stress concentrators, such as a hole, a steak or a groove, to validate the use of finite element software for obtaining this parameter in complex conditions represents a significant contribution to the design of mechanical elements, as it would allow to predict of a better way the stresses in the component parts of a machine due to the action of different load.

2. Conceptual framework

2.1. Stress concentrators

The stress concentrators are geometrical irregularities that cause an increase in the average effort that should be present in regions near these discontinuities, the relationship between the maximum stress that occurs and the average effort that should occurs is defined as stress concentration factor; which is determined by experimental or analytical methods and presented in graphical form for ease interpretation.

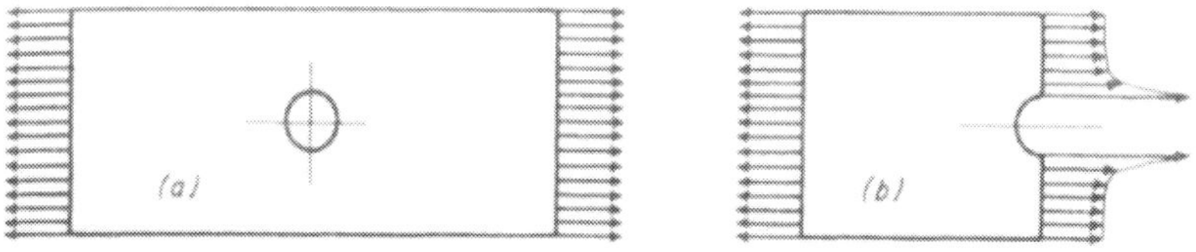

Figure 1. Stress distribution for a plate subjected to tensile load (a) away from the hole; (b) in the section of central hole [1].

A typical example of a stress concentrator is a rectangular bar with a central hole, subjected to tensile load, as shown in figure 1 (a). If the bar is cut in the cross section of the hole, the tensile stress will be as shown in figure 1 (b), the stress distribution along the cut surface is substantially uniform until reaching the vicinity of the hole, where efforts suddenly increase. The maximum value of stress at points is found by multiplying the average effort by a K factor of stress concentration [1, 2].

2.2. Stress concentration factors

The stress concentration factor for static load can be determined as the relation of the actual maximum real stress in the discontinuity and the average stress, and is obtained through the equation 1 [3].

$$K = \frac{Actual\ maximum\ stress}{Average\ stress} \qquad (1)$$

The average stress is determined through the basic equations and is defined according to the type of load that is acting on the element. In the case of an axial load that causes tension or compression, this value is calculated by equation 2 [4].

$$\sigma = \frac{Axial\ Force}{Cross\ section\ area} \qquad (2)$$

The stress concentration factors are the type of discontinuity on the geometry of the discontinuity and the type of load is experienced.

2.3. Finite elements

The general idea of the finite element method is the division of a continuous, in a smaller set of elements interconnected by a serie of points called nodes. The equations governing the behavior of continuous also govern the element. Of this way is achieved to pass from a continuous system (infinite degrees of freedom), which is governed by a differential equation or a system of differential equations to a system with a number of finite degrees of freedom whose behavior is modeled by a system of linear or non-linear equations [5, 6].

On nodes are materialized the fundamental unknowns of the problem; in the case of structural elements the fundamental unknowns are nodal displacements, and from these can be calculated the remaining unknowns of interest: stresses, deformations, etc. The finite element method and mathematical formulation is relatively new, although its basic structure has been known from long time, in recent years has undergone a great development because of advances in computer technology. These advances have been precisely those who have made available to the user many programs implementing finite element calculations. The correct handling of this type of program requires a deep knowledge not only of the material with which we work, but also the beginnings of the MEF; only in this case it will be able to ensure that the results of analyzes are in accordance with reality.

3. Methodology

The methodology for the development of the Project consisted of three phases:

3.1. Software simulations for determination of equivalent Von Misses stresses in sheets with central holes

The steps carried out for conducting the simulations were the following:

- Definition of the model that was used for simulation; in this case was defined a static model.
- Deposit the mechanical properties of the material to the database of the software and select it for the simulation; the data provided was: elastic modulus (E = 205 GPa), the yield strength in tension or compression (S_y = 250 MPa) and ultimate strength (S_u = 460 MPa).
- Realization of the graphic representation of the element to simulate; It was drawn a rectangular sheet of 0.00635 m thick; 0.0254 or 0.0508 m width and 0.4 m long. Additionally was performed 1 or 3 holes of different diameters and arrangements in the central zone.
- Realization of meshing process, the zones near discontinuities were refined.
- Definition of loads (magnitude and direction) and the supports that guarantee de static equilibrium.
- Perform the simulation and obtain the distribution of stresses present on the sheet.

3.2. Validation of the use of ANSYS software, in the obtainment of stress concentration factors

The software validation was done by comparing the data obtained from simulations performed on a sheet with a central hole and subjected to axial load; and the theoretical data found in the literature for the same conditions of stress concentration.

3.3. Determination of the graphics of stress concentrator factors in flat plates with three central holes, under axial load

The simulations to determine the stress concentration factor in flat plates with three central holes were performed for Wheel base holes of 0.015875 and 0.0254 m, and over a plate of 0.0508 m wide; 0.00635 m thick and 0.4 m length. The arrangement of the holes in the plate was made of two different forms, longitudinally and transversely; the diameters used were: 0.001016, 0.002032, 0.003048, 0.004064, 0.00508, 0.006096, 0.007112, 0.008128, 0.009144, 0.01016 y 0.011176 m; for the following loads: 5000, 10000, 15000, 20000N; these forces were applied on a timely basis in the centroid of the cross section of the plate.

4. Results obtained

4.1. Software validation

The results of the maximum stresses obtained from the simulations were used to determine the stress concentration factor in flat plates with a central hole, and subjected to axial load; these data are reflected in figure 2.

To validate the use of Ansys software in obtainment of concentration factors, we proceeded to determine the % error present between the theoretical data and those obtained through simulations, which are listed in figure 2; In addition was carried out an statistical analysis that showed a decision variable t of experimental student of 0.237, value that is below the 2.1448 obtained from the chart of this variable for a reliability of 95%, which allows us to affirm that there are no significant differences between the theoretical and simulated data.

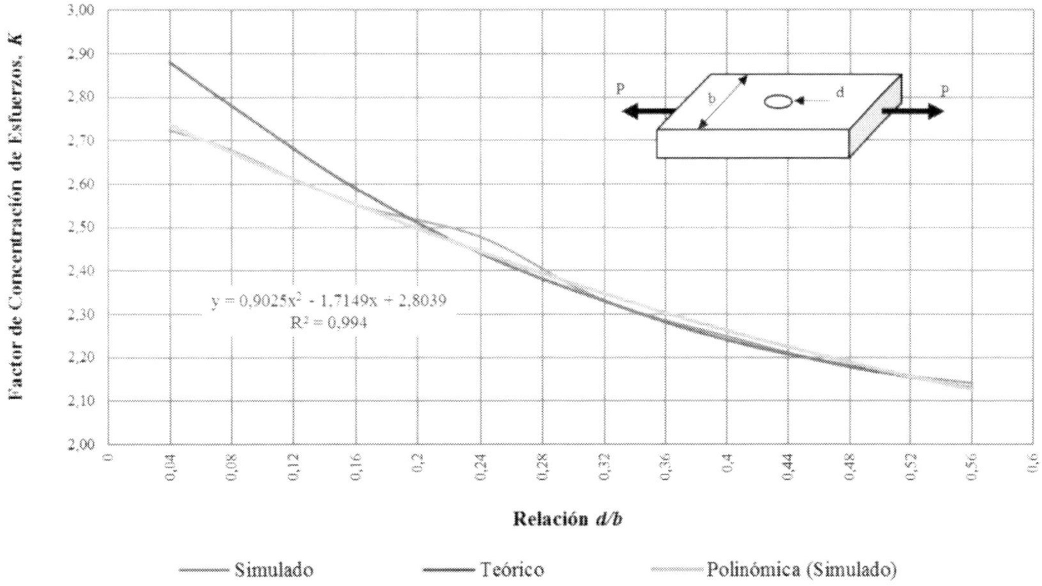

Figure 2. Comparison between the concentration factor data obtained through the simulation and the theoretical found in the bibliography.

4.2. Stress concentration factors in flat plates with three holes under axial load

The behavior of stress concentration factors of three holes located transversely on a flat plate, as expected, decreases as the relation between three times the diameter of the holes divided by the width of the plate increases; however this behavior doesn´t remain constant through the range but reaches a value which presents a inflection point and starts to increase, as shown in figure 3; this tendency may be due the weakening that suffers the plate in the zones near the upper and lower edges, and according with diameter of the hole increases, this region decreases and consequently the flow lines of stress in this area will be closer together.

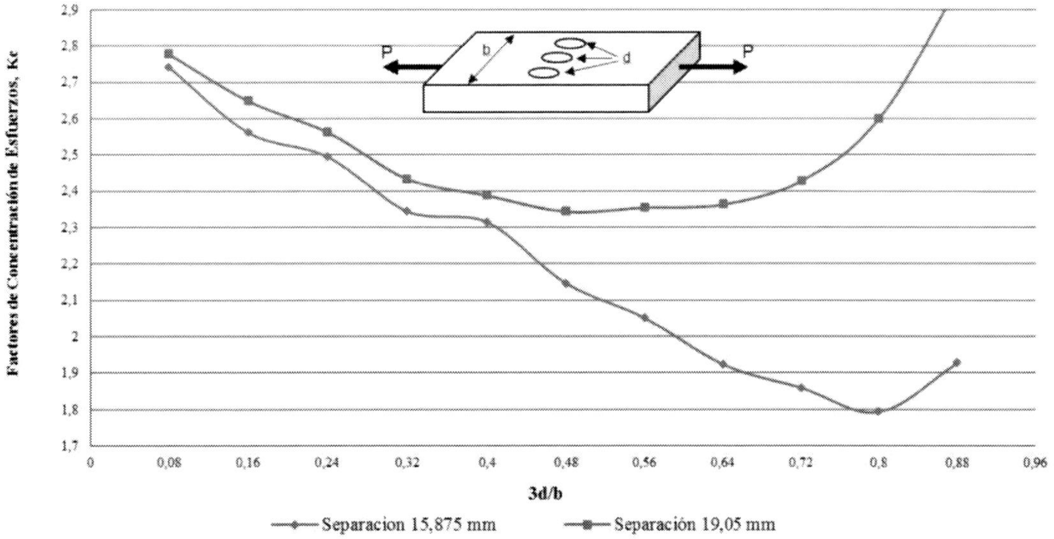

Figure 3. Stress concentration factors for a rectangular plate with three transverse holes.

The behavior of the stress concentration factor in three longitudinal holes located on a flat plate is shown in figure 4, as expected this parameter decreases as the relation between the diameter of the hole divided by the width of the plate increases; Also one can observe that the values of this parameter are independent of the distance between centers of the holes, because the data obtained are nearly equal for the two simulated separation conditions.

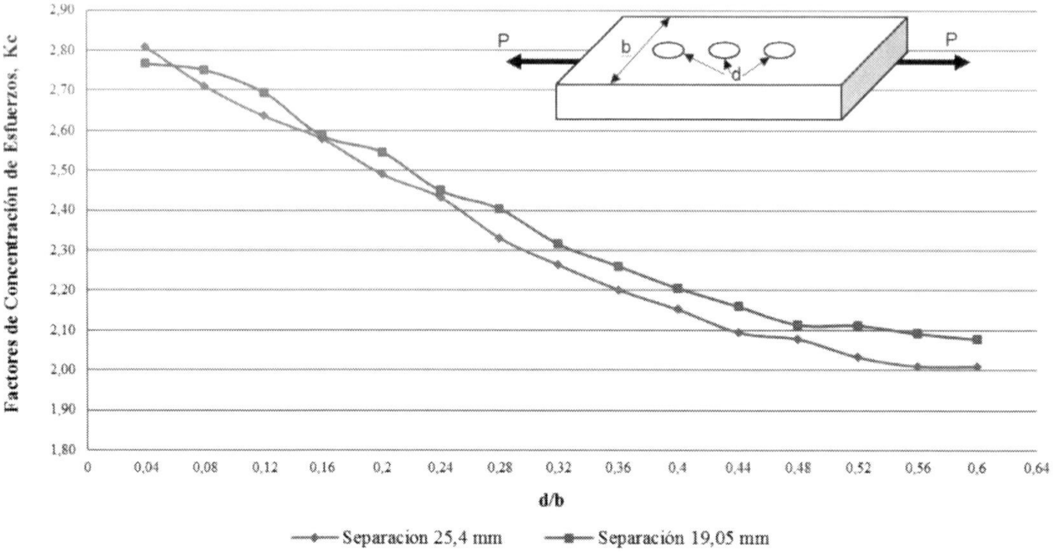

Figure 4. Stress concentration factors for a rectangular plate with three longitudinal holes.

5. Conclusions

The finite element software can be used as a tool to determine graphical stress concentration factors for flat plates with a central hole, and subjected to axial load, as could be statistically demonstrated that aren´t significant differences between the theoretical data and data obtained through simulation. The maximum calculated error between these data was 5.42% when geometric relationship between the hole diameter and the width of the plate is 0.004.

The stress concentration factors in flat plates with three holes located transversely and under axial load decreases when the relation 3r/d increases, which indicates that the larger the diameter of the hole the concentration factor is lower; but the maximum stress on the cross section increases. However, to ensure this correlation should be prevented that the hole diameter and the hole spacing exceeds a limit value, if this occurs the point of maximum stress changes the inner to the outer zone of the hole because the separation between the hole and the edge of the plate decreases significantly.

The stress concentration factors in flat plates with three holes located longitudinally and subjected to axial load decreases to the extent that the relation d/b increases and additionally the magnitude of the factor is unaffected by the distance between centers of the holes.

References

[1] M F Spotts and T E Shoup 1999 *Elementos de Máquinas 7ª ed.* (México: Prentice Hall)

[2] R G Budynas and J K Nisbett 2008 *Diseño en Ingeniería Mecánica de Shigley 8ª ed.* (México: McGraw-Hill)

[3] B J Hamrock, B O Jacobson and S R Schmid 2000 *Elementos de Máquinas* (México: McGraw-Hill)

[4] J M Gere 2006 *Mecánica de Materiales 6ª ed.* (Australia: Thomson)

[5] T R Chandrupatla and A D Belegundu 2002 *Introduction to Finite Elements in Engineering 3ª ed.* (New Jersey: Pearson Education)

[6] O C Zienkiewics 1982 *El Método de los Elementos Finitos* (Barcelona: Reverté)

Journal of Physics: Conference Series **466** (2013) 012036 doi:10.1088/1742-6596/466/1/012036

Design and construction of an impulse turbine

E Hernández[1]

[1] Grupo de Investigación en Desarrollo Tecnológico, Mecatrónica y Agroindustria – GIDETECHMA, Facultad de Ingeniería Mecánica, Universidad Pontificia Bolivariana seccional Bucaramanga, Floridablanca, Colombia.

E-mail: emil.hernandez@upb.edu.co

Abstract. Impulse turbine has been constructed to be used in the program of Hydraulic Machines, Faculty of Mechanical Engineering at the Universidad Pontificia Bolivariana, sede Bucaramanga. For construction of the impulse turbine (Pelton) detailed plans were drawn up taking into account the design and implementation of the fundamental equations of hydraulic turbomachinery. From the experimental data found maximum mechanical efficiency of 0.6 ± 0.03 for a water flow of 2.1 l/s. The maximum overall efficiency was 0.23 ± 0.02 for a water flow of 0.83 l/s. The design parameter used was a power of 1 kW, as flow regulator built a needle type regulator, which performed well, the model of the bucket or vane is built on a machine type CNC (Computer Numerical Control). For the construction of the impeller and blades was used aluminium because of chemical and physical characteristics and the casing was manufactured in acrylic.

1. Introduction

The Engineery Mechanic Programme at the Pontificia Bolivariana University was initiated in 2002, and a hydraulic machine is an important component of the curriculum which aims to provide an understanding of the physical principles involved in energy generation. To this end, the construction and test of an impulse turbine was devised to demonstrate the technology of Pelton turbines in hydropower plants.Hydroelectric generation plays an important role in global energy supply. For example, in 2010 it accounted for almost 16% of global electricity generation [1]. It's a renewable energy source because Earth´s water cycle is driven by solar energy. The Pelton turbine, which was patented by Lester Pelton in 1880, is commonly used in hydroelectric generation in geographical locations where high water head is available but volume flows are small.

The water flow is directed by one or more nozzles to strike the ridge of each vane tangentially in succession and in turn delivers the water´s kinetic energy in impulses. Therefore, the turbine is classified as an impulse turbine. In principle, the ridge acts to divide the water jet in order to achieve better mechanical efficiency. With an optimised jet and cup geometry, more than 90% of the power of the water jet can be transformed into mechanical power at the turbine shaft [2]. Since the Pelton wheel rotates in air, the action of the water and wheel rotation is easily observed by the experimenter. Direct observation of efficiency changes is an attractive characteristic from an educational perspective and it is said that Pelton himself got the idea for his turbine from such a direct observation [3].

2. Theory

The total extractable power of a water jet P_h can be found from [4]

$$P_h = \rho g h o \tag{1}$$

Where ρ is the density of water, g *the* acceleration due to gravity (9.81 m/s^2), h the available head of the water source and Q the volumetric water flow rate. The speed of the water jet C is found from the flow rate and the cross-section area A of a circular nozzle with diameter D

$$C = Q/A = 4Q/(\pi D^2), \tag{2}$$

The mechanical power P_a available at the turbine shaft can be determined by measuring the torque T on the shaft at a corresponding angular speed ω. The torque is found by measuring the tangential force F on a brake lever with moment arm length l (see figure 1).

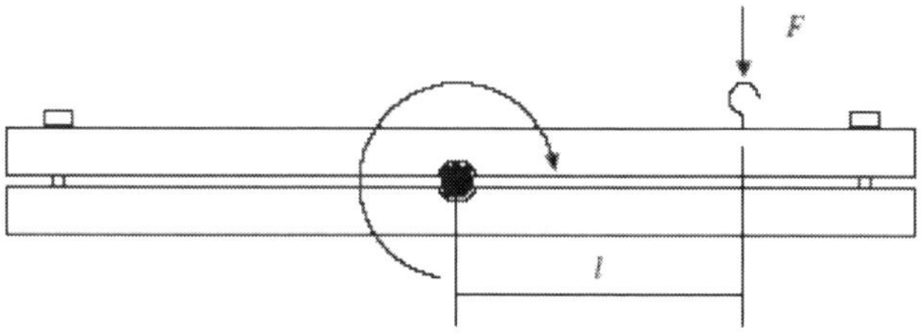

Figure 1. Measurement of the torque on the turbine shaft.

$$P_a = T\omega \tag{3}$$

The angular speed ω is equal to the tangential speed of the turbine U divided by the pitch radius R of the Pelton wheel

$$\omega = U / R \tag{4}$$

The mechanical efficiency of the turbine describes how effectively the available kinetic energy of the water jet is transformed into turbine motion and from (1) and (3) is

$$\eta_m = P_h / P_a \tag{5}$$

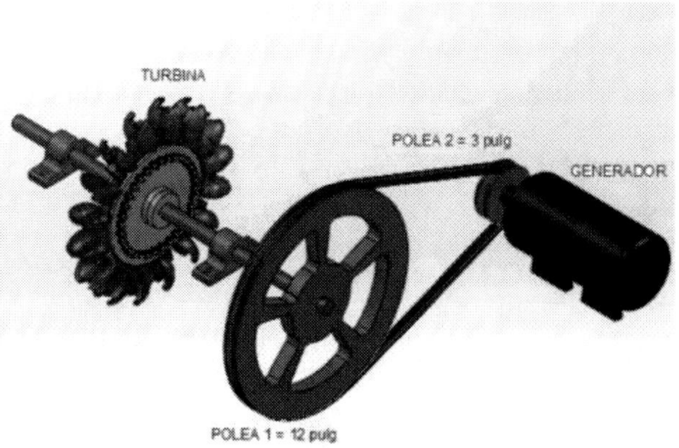

Figure 2. Pelton wheel water turbine experiment.

Cup and jet design are important parameters and it can be shown [5] that maximum theorical mechanical efficiency is achieved when the tangential speed of the turbine is roughly one-half the speed of the water coming from the jet or

$$U \approx 0.5\,C \tag{6}$$

When the turbine shaft is coupled to an electric generator which supplies electricity to a variable resistive load R (see figure 2), the electrical power P_e of the load is

$$P_e = IV \tag{7}$$

In which I is the load current and V the voltage across the load. The electrical efficiency η_e of the generator and the overall efficiency η_0 of hydroelectric generation are then

$$\eta_e = P_e/P_a \tag{8}$$

$$\eta_o = P_e/P_h \tag{9}$$

3. Methodology

The Pelton turbine consists of a cylindrical hub onto which 16 vanes (see figure 3) have been fastened using screws and right angle aluminium brackets. The hub and vanes were made of aluminium for easy machineability and resistance to the oxidation. The pitch radius of the turbine, which is the distance from the centre of the shaft to the centre of the impact point of the water jet (160 mm in this case), should be roughly 10 times the diameter of the jet (16 mm in this case) so that adjacent vanes do not interfere with the water flow [2]. For the turbine under discussion measured rotation speed have been in the range of 100 – 500 rpm with turbine power in the range of 20 – 150 W.

Figure 3. Pelton wheel.

Has been conducted experiments with the Pelton turbine using three different water volume flow rates, referred to as Q_1, Q_2 and Q_3. The water pressure p at the nozzle is measured for each Q value with the turbine shaft rotating freely.

The relationship between the tangential force on the shaft and the speed of rotation is then determined using the brake lever for each Q value. This relationship is generally found to be of a linear nature in the range of variables described (figure 3).

With the drive belt connected to the DC generator and the brake lever removed from the turbine shaft, the current and voltage of the resistive load are measured, while simultaneously measuring the rotational speed of the turbine. This is repeated for several values of load resistance. The measurement yields the electrical power produced by the generator as a function of rotational speed.

Using experimental data, the mechanical, electrical and overall efficiency of the hydroelectric plant are calculated for each Q value.

4. Results and discussion

Table 1 lists the Q values used, followed by the corresponding hydraulic power calculed from equation (1) using a net head available of 4 m. The corresponding maximum mechanical and electrical efficiencies are on the right side of the table along with the maximum total efficiency of the laboratory-scale hydroelectric plant.

The mechanical efficiency of the turbine (equation (5)) as a function of rotation speed is plotted in figure 4.

Table 1. The three Q values used in the experiment.

Q [l/s]	Ph [W]	Pa [W]	Pe [W]	η_m [W]	η_o
$Q_1 = 2.5$	98	44.1	19.6	0.45	0.20
$Q_2 = 2.1$	82.3	49.4	18.9	0.60	0.23
$Q_3 = 1.5$	58.8	29.4	8.82	0.50	0.15

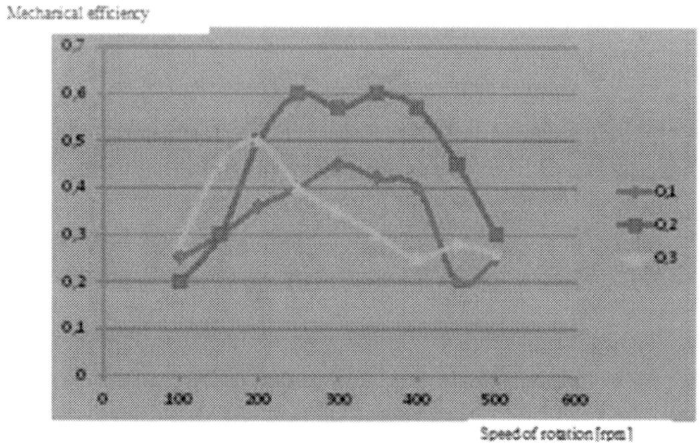

Figure 4. The mechanical efficiency of the Pelton turbine as a function of the speed of rotation.

5. Conclusion

A laboratory-scale Pelton turbine for hydroelectric generation has been constructed and used in the educational curriculum of the Engineering Mechanical Programme at the Pontificia Bolivariana University. From the experimental results, the turbine was found to have a maximum mechanical efficiency of 0.6 ± 0.03 for a flow rate of 2.1 l/s. The Pelton turbine demonstrate the principles of hydropower and are well suited in the education of students of Mechanical Engineering.

References

[1] International Energy Agency (IEA) 2010 *Key world energy statistics* (París: Soregraph)
[2] G Boyle 1996 *Renewable energy-power for a sustanaible future* (New York: Oxford Univresity Press Inc.)
[3] American Society of Mechanical Engineers 1991 *The Pelton water wheel collection. Grass Valley* (CA: North Star Power House Museum)
[4] F White 1994 *Fluid Mechanics 3rd ed.* (New York: McGraw-Hill Inc.)
[5] M El-Wakil 1984 *Power plant technology-international edition* (New York: McGraw Hill Inc.)
[6] P Fernández *Turbinas Hidráulicas* (Catabria: Departamento de Energía Electrica y Energética Universidad de Cantabria)
[7] C Mataix 1986 *Mecánica de fluidos y máquinas hidráulicas 2ed.* (Madrid: Ediciones del castillo S.A.)
[8] M Eisenring 1991 *Micro Pelton Turbines* (Germany: Swiss Center for appropriate technology)
[9] E Hernández 2006 *Handbook of statistics* (Bogotá: Educ.)
[10] P Encinas M 1983 *Turbomaquinas Hidraulicas Tercera Edición* (México: Editorial Limusa)
[11] D Agar, M Rasi 2008 *Renewable Energy* **33** 1517-22

Journal of Physics: Conference Series **466** (2013) 012037 doi:10.1088/1742-6596/466/1/012037

Preparation and characterization of masonry units, lightweight concrete based and agro-industrial wastes: a review

C X Díaz-Fuentes[1]

[1] Departamento de Arquitectura y Urbanismo, Universidad Francisco de Paula Santander – UFPS, Cúcuta, Colombia.

E-mail: xiodf@yahoo.com.br

Abstract. Discussion about the new composite materials that integrate agro industrial residues for the masonry unit's production, which are directed towards its implementation in projects of affordable housing, is a subject of interest to the public and productive sector of the country. For this reason, it presents a descriptive review of primary and secondary sources, which support the project under study. The methodology consisted in finding research articles in databases supported by the scientific community, which are ordered, integrated and prioritized, creating a matrix synthesis, which condensed the objectives, type of material, studied properties and main results found. It was found that the composite materials for masonry use mainly clay or cement as matrix and as reinforcement, agro waste like paper fibers, bamboo, rice husks, among others are used. Moreover, the properties that determine its potential use are low density, stress resistance and low thermal conductivity. Comparing the results with traditional specimens as the block of clay, concrete, adobe vs. experimental models made of the compounds analyzed, favorable results were obtained in the case of integrating waste materials into its composition, optimized their properties. Thus, science and architecture converge through recognition of the properties of materials that expand the alternatives of building spaces, economic and environmentally sustainable.

1. Introduction

Technological developments and the gradual progress in society have deeply impacted the environment becoming more visible its gradual decline coupled with increased energy consumption and emissions of the same product. According to the work done by Vargas Robles [1], the construction sector is one of the fields that degrades faster emerging territory hence the need to develop and use materials that are more easily integrated into the environment at the end its useful cycle [1-7, 11, 12, 15-22]. This paper seeks to publish the results of the development of two composites used as source added agro-industrial wastes. The results of such materials for possible use in masonry pieces where optimum performance is essential for its mechanical and thermal properties were analyzed [5-7].

Concerns about which are reflected in this study are based on discussions raised in the master of science and technology of materials and architectural plan of the University Francisco de Paula Santander where two aspects of great significance converge: material innovation [10] and the need to design a product that would facilitate the development of low-cost building systems for housing of priority and in emergency situations. All framed in constitutional rights of access to adequate housing [8-9].

Journal of Physics: Conference Series **466** (2013) 012037 doi:10.1088/1742-6596/466/1/012037

This work will allow the architecture curriculum form a line within a research group that plans to consolidate in the field of advanced materials applied to construction.

2. Materials and methods

The methodology used was a descriptive review of articles, books and works published in scientific grade which were obtained through databases such as Redalyc and Science Direct among others, the reliability criteria whereby it was determined that the items and the works selected were appropriate, at the same time were endorsed by the publishers of such committees and databases highly regarded in the scientific.

3. Results and discussion

It was reported that masonry units made from lightweight concrete matrices with reinforcements hydrated lime and alum as well as lightweight concrete, silica and paper sludge, showed low thermal conductivity, high mechanical strength and lower density than unreinforced units. [1-4]. that in accordance with ASTM C 168 (ASTM, 1995,s.p), [11, 15, 16] over the standard terminology related to thermal insulating materials, where the cellulosic fibers act as insulators, usually derived from paper, cardboard or wood with or without binders for bonding. Thus as required by the ASHRAE [14], as the studied composites may be in the following classifications: a) As insulating cement, which can be a mixture of binder, reinforcing and dry additives, which when incorporating water could be applied in wet surfaces; b) As rigid material, by its possibility to be fabricated in block, panel or sheet.

Masonry samples prepared from clay and rice hulls and lightweight concrete specimens and paper sludge (1 to 5.22), reported a significant decrease in weight due to lower density of the material; in addition when comparing with the properties of common brick, block and concrete block adobe, its qualities were superior to yield better behavior by low thermal conductivity which places them at an advantage versus traditional materials (See table 1 and 2), obtaining as added value the reuse of agroindustry residuals which affect reducing costs and environmental impact.

Table 1. Density and thermal properties of brick adobe and silica mixture + paper sludge[a].

Properties	Brick	Adobe	Silica-Paper Sludge	Concrete Block
Density (kg/m3)	1800	1620	855	2300
Thermal conductivity (W/m °C)	0.72	0.49	0.383	1.08
Specific heat (J/kg °C)	829	1240	1513	920
Thermal diffusivity (m^2/s)	*1.7x10-3	*8.7x10-4	1.02x10-3	1.84x10-3
Thermal Delay (h)	*4h 38'	*6h 32'	6h 28'	4h 31'
Thermal buffer (%)	*30	*18	18	*30
Thermal resistance (m^2 °C/W)	*0.208	*0.306	0.391	0.138

[a] Source [2].

Table 2. Compressive strength and density of local materials[a].

Material	Mechanical Strength (Kg/cm^2)	Density (Kg/m^3)
Common Brick	18	1800
Adobe	8	1620
Silica + paper sludge	34	855
Block	40	2100

[a] Source [2].

4. Conclusions and recommendations

It was shown that there are now growing agro-industrial waste reuses as fly ashes, paper sludge, bamboo, rice husk, silica and others. Thus is raised the constant need for alternatives to conventional materials, offering opportunities in terms of their characteristics, and properties that allow the applicability in the design and construction of living spaces.

It was evident a significant cost reduction due to the utilization of industrial wastes as aggregate in foundation for its low volumetric mass and energy consumption for its artificial acclimatization.

It further reported that Latin American countries like Mexico and Brazil are leading such research while in Colombia disclosure is still nascent.

It was evidenced that these studies are not made of materials analysis using techniques such as x-ray diffraction, scanning electron crystallography, which would allow a better understanding of the internal structure of the material.

These are used most commonly property evaluations known as physical mechanical strength, thermal and acoustic conductivity.

References

[1] Vargas Robles, Luis Roberto 2007 *Uso de fibras de papel periódico, cal hidratada y alumbre, en la elaboración de un material compuesto* (México: Universidad de Colima)
[2] Bojórquez G, Luna A, Gallegos R 2000 *Temas de Ciencia y Tecnología* **10** 3
[3] Bojórquez G 1999 *Lodo de papel como agregado en un concreto ligero para bloques de mampostería* (México: Universidad Autónoma de Baja California)
[4] Bojórquez G 2005 *Desarrollo de un nuevo material a base de fibras orgánicas del sudeste mexicano en matriz de cemento Pórtland para producir paneles ligeros* (Cuba: Centro de estudios de construcción y arquitectura tropical, Instituto superior politécnico José Antonio Echeverría)
[5] D Eliche-Quesada, C Martínez García, M L Martínez Cartas, M T Cortes Palominok, L Pérez Villarejo, N Cruzpérez, F A Corpas Iglesias 2011 *Elsevier B.V. Applied Clay Science* **52**.
[6] M Ahmaruzzaman 2010 *Progress in energy and combustión science* **36**.
[7] M D Safiuddin, Mohd Zamin Jumaat, M A Salam, M S Islam and R Hashim 2010 *Journal of the Physical Sciences* **5-13** 1952
[8] Torres Tovar, Carlos Alberto 2010 *Revista Bitácora Urbano Territorial* **17-2** 7
[9] Villar Lozano, Mayerly Rosa 2009 *Revista de Arquitectura* **11** 4
[10] Brostow W 1981 *Introducción a la ciencia de los materiales* (México: Editorial Limusa S.A.)
[11] Marúm E 1989 *La producción de celulosa y papel en México* (México: Universidad de Guadalajara)
[12] Vinson J Y, Sierakowski R 1987 *The Behavior of Structures Composed of Composite Materials* (Paises Bajos: Kluwer Academic Publishers)
[13] American Society for Testing and Materials 1995 *Annual book of ASTM standards* (Estates Unites:)
[14] American Society of Heating, Refrigerating and Air Conditioning Engineers 2001 *Ashrae Handbook* (Estates Unites)
[15] Lo Tommy, Cui H Z, Leung H C 2007 *Materials letter* **58** 2595
[16] Ghavami Khosrow 2004 *Journal Cement and Concrete Composite* **26** 1
[17] García J and Silva G 2011 http://www.conarquitectura.com/articulos%20tecnicos%20pdf/16.pdf
[18] Larubia M 2011 www.ehu.es/sem/macla_pdf/macla13/Macla13_135.pdf
[19] Arcitex 2009 *Reutilización de residuos textiles en la industria de la arcilla cocida* (España: Memoria Aitemin)
[20] Quaranta N 2011 http://uninorte.edu.co/divisiones/Ingenierias/IDS/upload/File/Memorias%20II-SIIR/3b-Quaranta-Argentina-001.pdf.
[21] Balaguera L C and Carvajal J A 2004 *Estudio para producir bloque aligerado a partir de mezclas de arcilla, cenizas volantes y poliestireno expandido en la empresa cerámicas Támesis* (San José de Cúcuta: Universidad Francisco de Paula Santander)
[22] Andrés Mafla B 2009 *Inventum* **6** 74

Journal of Physics: Conference Series **466** (2013) 012038 doi:10.1088/1742-6596/466/1/012038

Characterization of materials formed by rice husk for construction

A M Portillo-Rodríguez[1]
[1] Aspiring Master Materials Science and Technology UFPS-Colombia

E-mail: asportilloro@hotmail.com

Abstract. This review article delves into the use of agro-industrial wastes, which in construction field provides alternatives for environmental problems with the use of them. This fact enables development and lower costs for new options in the brick, cluster, mortar and concrete industry, what represents benefits for environment, housing and generally everything related to construction, looking for sustainability. For that reason a literature review is made to support the theme focusing on the use of rice husk in its natural, ground or ash state for manufacturing elements with clay masonry, precast and optimization of concrete and mortars. The technique used is based on scientific articles and researches found in reliable databases that were analyzed and integrated into a synthesized structure, which summarized the objectives, analysis processes, the physical and mechanical properties and finally the results. The conclusions are focused on potentiality of elements production in the construction development based on the high effectiveness like thermal insulation, low density and various benefits offered by high silica content pozzolanic properties, etc.

1. Introduction

The excessive global population growth brings high demand for buildings with higher levels of comfort and to low costs, but it also appears the need to produce sustainable architecture in order to reduce energy consumption while lowering environmental impact. Around this, a worldwide interest in optimizing energy resources is growing as well as has the primary responsibility to promote comfort, it is also the most expensive component [6], therefore this brings the need for alternatives comprehensive address bioclimatic, economic aspects, material substitution technologies and environmentally friendly [10, 15].

Against this background it explores the use of rice husk that besides of being an agro-industrial by-product of big generation and accumulation of large producing areas around the world, it is emerging as an important alternative in the development of composite materials for possessing important features requirements for buildings such as thermal isolation, sound isolation, low density [5] force [15], porosity, in addition to its high potential as a source of SiO_2, which by their pozzolanic properties for strength and durability, important aspects in the manufacture of bricks, cement industry, paints, resins and plastics, among other applications [14, 18, 19, 21, 16].

This paper reviews the research on rice husk and its use in composite materials for construction, to provide an analysis of the results that will highlight the aspects studied and the results to guide future extensions and innovations in order to optimize processes and materials in the search for new applications and possibilities for development.

At Francisco de Paula Santander University it is generated a meeting point between architecture and materials science that provides wide space for common concerns, aimed at generating interdisciplinary exploration of new materials to help solve common situations needs of big part of population in search of low cost buildings, to optimize the characteristics of traditional materials, with ease of construction, and also with a reduced environmental impact.

2. Methods

The technique used for the development of this work lay in the review of scientific studies on rice husk oriented construction field in different countries of the world in the last 10 years, which were obtained in Science Direct, Redalyc, Publindex and Scielo, and other prestigious databases in science, determining factor in the definition of reliability criteria and inclusion of the results analyzed and synthesized.

3. Results and discussion

The results show that the development of composites with rice husk in different presentations is useful as an alternative to conventional materials substituent in construction and contributes to reducing the environmental impact caused by the waste products and of the respective burning [15].

It is reported that replacement of cement in a certain quantity by rice husk ash, primarily with chemical treatment improves the concrete, mortar and grout Portland cement, the characteristics of resistance to compression (see figure 1) [4] and flexural strength, durability and reduction of permeability [7, 11, 13, 14], appearance determined by the high content of amorphous SiO_2.

The obtained results show that the structure in concrete and mortars with added clay hull ash is different according to the fineness of the ash to be incorporated, where mixtures of rice husk ash ground are more homogeneous and compact for its amorphous state, as opposed to blends with rice husk ash naturally occur more open structures, and in addition to variables such as temperature and time, show changes in characteristics such as workability, compressive strength, water permeability and porosity [11, 12, 13, 18, 19].

Unlike the rice hull ash, the rice hull inclusion without incinerating in cement mixtures has disadvantages by its porosity, hygroscopy and organic content, which brings the water demand of the pulp segregation and consequently decreased strength [5]. But the use with or without pretreatment may provide interesting results in the incorporation of rice husk fibers in matrix composites with concrete, clay or synthetic non-structural applications because it provides significant characteristics of low density and high porosity that prominently located to produce building materials with thermal insulation properties (see table 1) and soundproofing. It is recorded that a greater amount of rice husk is obtained heat absorption [6], but their behavior is reversed related to particle size, which increases its conductivity by reducing its size [1].

The possibilities are evident that rice husk provides according to the various states in which they present, and in its natural state, ground, ashes, ground into ashes, ashes with different treatments and generally in different combinations; and according to this, it provides a variety of options for improving characteristics matrix composites with clay, cement or synthetic agglomerates, such as low thermal conductivity, low density, durability and resistance to degradation [1, 2, 5, 15, 17, 22, 24].

The results show that the main variables in determining efficiency and application possibilities of the material are determined at the process temperature, the heat treatment time and the presentation and the percentage of inclusion of rice hulls [1, 15].

There has been research that lead to generating more energy potential environmentally friendly sustainable obtaining rice husk ash, from the idea of taking advantage of the process in an integral way. The heat produced in combustion creates thermal energy that can be harnessed in different production processes, capturing flue gases including carbon dioxide (CO_2), Calcium Carbonate ($CaCO_3$) and other agricultural uses [9, 10, 23, 20, 21].

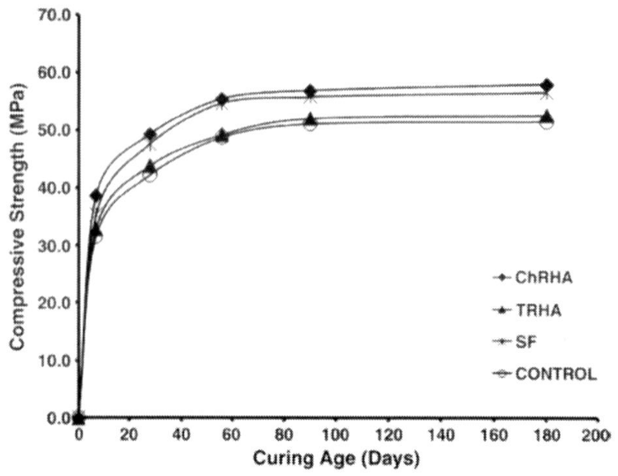

Figure 1. Compressive strength of concrete mixes. Source: [4].

Table 1. Comparison of thermal conductivity of rice husk and some commercial insulation[a].

Material	K (W/m*k)
Mineral wool	0.0303
Polystyrene	0.03030
Rice husk	0.0360
Fiberglass	0.0380
Agglomerated cork	0.0450

[a] Source [22].

4. Conclusions and recommendations

This work shows that research is defined cost reduction versus conventional products, which determines that the use of rice husk provides many options in the field of construction, and well outlined the possibility to meet the demand in and increase accessibility to the entire population [7, 14, 16].

Materials with concrete, clay and polymer may be produced by using rice husk to satisfy needs durability, thermal and acoustic insulation, and low cost, even if not fulfill high quality standards in resistance, but possible for smaller buildings specifications [1, 6, 15,].

It is clear that rice husk has drawbacks such as high organic matter content and moisture content, which requires treatment and controlled processes to optimize results, which provides that for the development of rice husk, technology and technical feasibility studies, economic and environmental, mainly in the case of the ashes should be introduced because the burning and grinding processes are major sources of pollution [9, 14].

References

[1] Gökhan G and Osman S 2013 *Construction and Building Materials ScienceDirect* vol **40** 390
[2] Cruz J C, González J, Perdomo L, Rodríguez P 2012 *Ingeniería Mecánica vol* **15** 115
[3] Calero F and Vásconez L 2012 *Scientific Articles – Carer Mechanical Engineering*
[4] Salas A, Delvasto S, Mejía de Gutiérrez R and Lange D 2009 *Cement and Concrete Research* **39** 773
[5] Serrano T, Borrachero M, Monzó J and Paya J 2012 *Dyna rev.fac.mynas* **79**.

[6] Sisman C, Gezer and Kocaman I 2011 *Bulgarian Journal of Agricultural Science* **17** 40
[7] Memon S, Shaikh S and Akbar H 2011 *ScienceDirect. Construction and Building Materials* **25** 1044
[8] Pineda P, Bedoya C and Rosales A 2011 *Dyna* **165** 207
[9] Prada A and Cortes C 2012 *Orinoquia* **14** 155
[10] Martinez J, Pineda T, López J and Betancurt M 2010 *Revista Facultad Ingeniería Universidad de Antioquia* **51** 97
[11] Alireza N, Suraya A, Fara H and Mohamad M 2010 *ScienceDirect. Construction and Building Materials* **24** 2145
[12] Batic O, Giaccio G, Zerbino R and Isaia G 2010 *VI International Congress on Pathology and Recovery Structures* (Argentina: Córdoba)
[13] Rashid M, Molla K and Ahmed T 2010 *European Journal of Scientific Research* **40** 471
[14] Mafla A 2009 *Inventum* **6** 74-78
[15] Bedoya C, Pineda P and Rosales A 2010 *Ingeniería y Ciencia* **5** 155
[16] Behak L, Peres W 2008 *Revista Ingeniería de Construcción* **23** 34
[17] Aguila I and Sosa M 2008 *Revista de la Facultad de Ingeniería* **23** 55
[18] Solarte S, Ospina M, Aperador W and Mejia de Gutierrez R 2007 *Scientia Et Technica* **XIII** 443
[19] Arcos C, Macías D and Rodríguez J 2007 *Revista Facultad de Ingeniería Universidad de Antioquia* **41** 7
[20] Valverde A, Sarria B and Monteagudo J 2007 *Scientia Et Technica* **XIII** 255
[21] Ahumada L and Rodríguez J 2006 *Rev. Acad. Colomb. Cienc.* **30** 581
[22] Cadena C and Bula A 2002 *Ingeniería y Desarrollo* **12** 1

AUTHOR INDEX